905

UNPOPULAR ESSAYS
on Technological Progress

UNPOPULAR ESSAYS
on Technological Progress

NICHOLAS RESCHER

University of Pittsburgh Press

Published by the University of Pittsburgh Press, Pittsburgh, Pa. 15260
Copyright © 1980, University of Pittsburgh Press
All rights reserved
Feffer and Simons, Inc., London
Manufactured in the United States of America

Library of Congress Cataloging in Publication Data

Rescher, Nicholas.
  Unpopular essays on technological progress.

  Includes index.
  1. Technology—Philosophy. 2. Science—
Philosophy. I. Title.
T14.R42   601' .9   79-21648
ISBN 0-8229-3411-6

"The Environmental Crisis and the Quality of Life," © 1974 by the University of
Georgia Press, is reprinted from *Philosophy and Environmental Crisis,* ed.
William T. Blackstone, by permission of the University of Georgia Press. "The
Allocation of Exotic Medical Life Saving Therapy," © 1969 by the University of
Chicago, is reprinted from *Ethics* 29 (1969). "Ethical Issues Regarding the
Delivery of Health Care" first appeared in *Connecticut Medicine* 41 (1977).
"Economics Versus Moral Philosophy: The Pareto Principle as a Case Study"
first appeared in *Theory and Decision* 10 (1979), 169–79, © 1979 by D.
Reidel Publishing Company, Dordrecht, Holland; reprinted by permission of
D. Reidel Company.

For Catherine
patris amore

Contents

Preface

Notwithstanding their topical diversity, the essays brought together in this volume share a common theme: the impact for human concerns of the complexity and scope of modern technological progress and its economic and social ramifications. Most of these essays grew out of invitations to contribute a paper to some sort of conference or symposium. Despite the professional's predilection for shoptalk with colleagues, a philosopher even nowadays finds himself impelled by various pulls and pressures to enter into the nontechnical discussion of timely issues of the day.

Some of these essays have appeared before, though in each case they have been revised for publication. Chapter 2 is a revised version of a paper presented in a symposium, Philosophy and Environmental Crisis, at the University of Georgia in Athens in February 1971 and initially published in William T. Blackstone, ed., *Philosophy and Environmental Crisis* (Athens, Georgia, 1974). Chapter 3 is a revised version of an essay first published in *Ethics* 29 (1969). Chapter 4 has been revised from a paper read at a conference on Ethical Issues in the Distribution of Health Care at Brown University in April 1975 and subsequently published in *Connecticut Medicine* 41 (1977). Chapter 6 is a revised version of a paper published in *Theory and Decision* 10 (1979). I wish to thank the editors of these publications for their kind permission to include this material in the present volume.

Chapter 1 was prepared as a Phi Beta Kappa Lecture delivered at the University of Pittsburgh, December 6, 1977. Chapter 5 is based on a paper read in a series of lectures on Justice in America at the University of Vermont in March 1977. Chapter 8 grew out of a lecture delivered at the Sixteenth World Congress of Philosophy held at Düsseldorf in 1978.

I want to thank Virginia Chestek for her able assistance in preparing this material for publication and in helping me to see it through the press.

UNPOPULAR ESSAYS
on Technological Progress

Technological Progress and Human Happiness

1

Does increased knowledge of nature, and technological mastery over it, enhance man's happiness and satisfaction, or is what we honorifically, nay almost reverently, characterize as "progress" really irrelevant to this central issue regarding the human condition? This question lies at the dead center of any examination of the relationship between technology and humanistic concerns. It goes to the very heart of the matter: the linkage between man's knowledge and "mastery over nature" on the one side, and his humane life-world of thought and feeling on the other. It is a question that theoreticians of science and technology generally ignore. But humanists have often touched on it—it may be viewed as a key issue in Goethe's *Faust,* for example.

Writing in 1920, the able British historian of progress J. B. Bury painted the following picture: "The very increase of 'material ease' seemed unavoidably to involve conditions inconsistent with universal happiness; and the communications which linked the peoples of the world together modified the methods of warfare instead of bringing peace. . . . [The modern triumphs of the advance of man's aims] hardly seemed to endanger the conclusion that, while knowledge is indefinitely progressive, there is no good reason for sanguine hopes that man is 'perfectible' or that universal happiness is attainable."[1] This quote provides an appropriate setting for deliberating about the implications of the impressive modern growth in our technological competence for human happiness and the tenor of the condition of man.

The Historical Dimension

Let us look briefly at the historical dimension of this issue. The question of the reality and significance of progress has been de-

3

bated since the quarrel between the ancients and the moderns regarding the relative importance of the wisdom of classical antiquity as compared with modern learning was launched in the late Renaissance. At the dawn of modern science in the seventeenth century, the leading figures from Bacon to Leibniz all took a highly optimistic view. Man's knowlege was about to enter a new era, and his circumstances and conditions of life would become transformed in consequence. Consider a typical passage from Leibniz:

I believe that one of the biggest reasons for this negligence [of science and its applications] is the despair of improving matters and the very bad opinion entertained of human nature. . . . But . . . would it not be fitting at least to make a trial of our power before despairing of success? Do we not see every day new discoveries not only in the arts but also in science and in medicine? Why should it not be possible to secure some considerable relief from our troubles? I shall be told that so many centuries had worked fruitlessly. But considering the matter more closely, we see that the majority of those who dealt with the sciences have simply copied from one another or amused themselves. It is almost a disgrace to nature that so few have truly worked to make discoveries; we owe nearly everything we know . . . to a handful of persons. . . . I do believe that if a great Monarch would make some powerful effort, or if a considerable number of individuals of ability were freed from other concerns to take up the required labor, that we could make great progress in a short time, and even enjoy the fruits of our labors ourselves.[2]

Such a perspective typifies the seventeenth-century view of the potential of scientific and technical progress for making rapid and substantial improvements on the human condition.

By the nineteenth century the bloom of ameliorative hopefulness was definitely beginning to fade. The lines of thought worked by Malthus and Darwin introduced a new element of competition, struggle, and the pressure of man against man in rivalry for the bounties of nature. The idea that scientific and technological progress would result in enhanced human satisfaction/contentment/happiness came to be seriously questioned. Writing about 1860, the shrewd

German philosopher Hermann Lotze said: "The innumerable individual steps of progress in knowledge and capability which have unquestionably been made as regards this production and management of external goods, have as yet by no means become combined so as to form a general advance in the happiness of life. . . . Each step of progress with the increase of strength it brings, brings also a corresponding increase of pressure."[3] Thus already over a century ago, thoughtful minds were beginning to doubt that man's technical progress offers him a royal route to happiness.

Some Distinctions

Before turning to a closer exploration of this issue, let one or two important preliminary points be settled. For one thing, it is necessary to approach the issue of the human advantageousness of technical progress via the important distinction between negative and positive benefits. A negative benefit is the removal or diminution of something bad. (It is illustrated in caricature by the story of the man who liked to knock his head against the wall because it felt so good when he stopped.) A positive benefit, on the other hand, is one which involves something that is good in its own right rather than by way of contrast with an unfortunate predecessor.

Now there is no doubt that the state of human well-being has been, or can be, enormously improved by science and technology as regards negative benefits. There can be no question but that technical progress has enormously reduced human misery and suffering. Consider a few instances: medicine (the prevention of childhood diseases through innoculation, anesthetics, plastic and restorative surgery, hygiene, dentistry, etc.); waste disposal and sanitation; temperature control (heating and air conditioning). It would be easy to multiply examples of this sort many times over.

But the key fact remains that such diminutions of the bad do not add up to a good; the lessening of suffering and discomfort does not produce a positive condition like pleasure or joy or happiness. Pleasure is not the mere absence of pain, nor joy the absence of sorrow. The removal of the negative does not create a positive — though, to be sure, it abolishes an obstacle in the way of positivity.

And so the immense potential of modern science and technology for the alleviation of suffering does not automatically qualify it as a fountain of happiness.

Moreover, in various ways science and technology have created a setting for life which is counterproductive for the achievement of happiness. One instance is modern military technology and life under the shadow of the atomic sword; another is the overcrowding of human populations, the product largely of modern medical, hygienic, and agricultural technology. There is organizational centralization that has put all of us at risk as victims of disgruntled employees, irate consumers, disaffected citizens, political terrorists, and other devotees of direct action against innocent bystanders as a means to the realization of their own objectives. This list is easily prolonged, but here I want to make a larger and perhaps foolhardy assumption. For I am going to adopt the somewhat optimistic stance that all the problems that science and technology have created, science and technology can also resolve. And accordingly I am going to leave this negative aspect of the situation wholly out of account and look at the situation if best comes to best, so to speak.

The question to be considered is, Even if we view the consequences of science and technology for the human condition in the most rosy light and look on them in their most favorable setting (not in Ethiopia, say, or India, but in the United States and Western Europe where the most advantageous and least problematic conditions have prevailed), is it really clear that science and technology have wrought benign effects upon the condition of human happiness, viewed in its positive aspect?

Transition: The Sociological Perspective

Let us now move off in a different direction. The issue today has a new dimension. In the past, discussion has proceeded on a speculative basis, and the participants were principally philosophers and philosophically inclined historians or students of social affairs. But now the sociologists and social psychologists have taken over. It thus becomes possible to bring statistical data to bear and to look at the empirical situation. We need no longer speculate about the relation between progress and happiness. We can "go out into the field" and find out how things go in what tough-minded social scientists like to refer to as "the real world." That is, we can proceed by means of questionnaires and the whole paraphernalia of empirical social science. When we do this, all but the most cynical among us are in for some surprises.

The Negative Correlation Between Progress and Perceived Happiness

If the thesis that increased physical well-being brings increased happiness were correct, one would certainly expect Americans to regard themselves as substantially happier today than ever before. This expectation is not realized.

A substantial body of questionnaire data makes possible a survey of trends in the self-evaluated happiness of Americans. Operating with increasing sophistication, various polling organizations have queried massive numbers of representative Americans as to their degree of happiness: whether "very happy," "fairly happy," "not happy," or the usual "don't know." Some relevant findings are set out in schematic form in table 1.1 and in figure 1.1. There is doubtless some looseness in the comparison of these data collected by somewhat different procedures by different organizations,[4] but a relatively clear and meaningful picture emerges all the same: There has been no marked and significant increase in the self-perceived happiness of Americans to accompany the very substantial rise in the standard of living that has been achieved in the postwar period.

Moreover, for about a generation now sociologists and social psychologists have gone about asking people for their judgments regarding the changing condition of human happiness (contentment, satisfaction) in the face of a steadily rising standard of living. Consider some illustrations:

Science has made many changes in the way people live today as compared with the way they lived fifty years ago. On the whole, do you think people are happier than they were fifty years ago because of those changes, or not as happy? (Roper/Minnesota, 1955)

Happier	Not as Happy	No Difference	No Opinion/ Other
36%	47%	15%	3%

Thinking of life today compared to back when your parents were about your age — do you think people today generally have more to worry about, or that there's not much difference? (Roper/Minnesota, 1963)

More Today	Less Today	No Difference	Don't Know/ No Answer
68%	8%	20%	4%

Figure 1.1: Happiness of Americans by Income Category

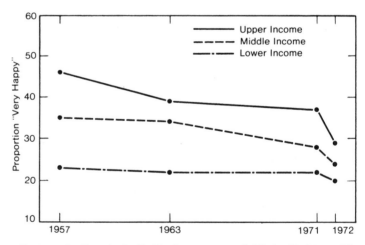

Source: A. Campbell, P. E. Converse, and W. L. Rodgers, *The Quality of American Life: Perceptions, Evaluations, and Satisfactions* (New York, 1975).

The same sort of result comes from a 1971 study by the Institute of Social Research:

Are things getting better in this country (USA)?

Better	Worse	About the Same
17%	36%	47%

This, of course, stems from 1971 — before Watergate, before the spate of "shortages" (gasoline, heating fuel, sugar, toilet paper, etc.), and before the period of major inflation and the whole complex of our latter-day discontents.

Such data indicate a clear result on the question, Does progress

Table 1.1: Self-classification of Americans in Point of Happiness

Year and Organization	Very Happy	Fairly Happy	Not Happy	Don't Know	Score[a]
1946 (AIPO)	39%	50%	9%	2%	110
1947 (AIPO)	38	57	4	1	125
1949 (AIPO)	43	44	12	1	106
1957 (SRC)	35	54	11	—	102
1963 (NORC)	32	51	16	—	83
1965 (NORC)	30	53	17	—	79

Sources: for 1946–1949: Hazel Erskine, "The Polls: Some Thoughts About Life and People," *Public Opinion Quarterly* 28, no. 3 (Fall 1964); for 1957: Gerald Gurin, Joseph Veroff, and Sheila Feld, *Americans View Their Mental Health* (New York, 1960), p. 22; for 1963 and 1965: Norman M. Bradburn, *The Structure of Psychological Well-Being* (Chicago, 1969), chap. 3, table 3.1.

Note: This table is compiled from the results of questionnaire studies conducted by AIPO (American Institute of Public Opinion, Princeton, N. J. — the Gallup organization), SRC (Survey Research Center, University of Michigan), and NORC (National Opinion Research Center, University of Chicago).

a. In computing the "score," we set the following figures: very happy = +2, fairly happy = +1, not happy = –2, and don't know = 0.

enhance happiness? When we approach the issue in this way, in terms of people's perceptions, the answer is emphatically negative. Half the people or more apparently think that the current hedonic quality of people's lives bears ill comparison with earlier stages of the "march of progress."

It is of interest to view such findings in the light of more detailed questionnaire studies, such as the following:

Do you think the human race is getting better or worse from the standpoint of health? knowlege? inner happiness? peace of mind? (AIPO, 1949)

	Better	Worse	No Difference	No Opinion
Health	73%	18%	6%	3%
Knowledge	82	7	7	4
Inner happiness	21	51	18	10
Peace of mind	17	62	11	10
Peace of mind by education				
College	16	74	6	4
High school	18	63	11	8
Grade school	17	57	13	13

Note: For comparable and supporting data see Hadley Cantril and Mildred Strunk, Public Opinion: 1935–1946 (Princeton, N.J., 1951), p. 280.

The contrast is a striking one here. Substantial majorities envisage a course of substantial improvement in terms of material and intellectual attainments. Nevertheless sizable pluralities take the view that our situation is deteriorating as regards "inner happiness" or "peace of mind," and, interestingly enough, the more highly educated the group being sampled, the more emphatic this sentiment becomes. The results of such surveys indicate that in fact a substantial majority of people believe there is a negative correlation between progress and happiness.[5]

Such evidence, to be sure, relates to the subjective impression of the people interviewed.[6] But there are also relevant data of a more objective kind that indicate a failure of Americans to achieve a higher plateau of personal happiness in the wake of substantial progress in the area of social welfare. For one thing, the suicide rate per 100,000 population per annum has hovered with remarkable stability at about eleven, plus or minus one-half, ever since World War II. Moreover, since 1945 a steadily increasing number of Americans are being admitted to mental hospitals, and, on the average, are spending an increasingly long time there. Statistical indicators of this sort are readily matched by a vast body of psychiatric data. Even political observers, who certainly have their hand on the nation's pulse, have become concerned over our inability to translate augmented personal affluence into increases in happiness. Richard Nixon in his first State of the Union message said: "Never has a nation seemed to have had more and enjoyed less."[7]

The Preference for the Present

Given such extensive—and continuing—indications that happiness is on the wane, it would seem clear that people would hanker after "the good old days," that many or most people would prefer to have lived in a bygone, happier time.

So, indeed, it might well appear. But the statistics obtained in the field indicate that this expectation is altogether wrong.

Do you think you would have rather lived during the horse-and-buggy days instead of now? (Roper, 1939)

Yes	No	No Opinion
25%	70%	5%

If you had the choice, would you have preferred to live in the "good old days" rather than the present period? (Roper/Minnesota, 1956)

Yes	No	Other
15%	57%	29%

Here we have something of a paradox. On the one hand, people incline to believe that "things are going to the dogs"; on the other hand, people evince no real preference for "the good old days." And these findings are altogether typical of findings obtained over the past generation. Invariably, Americans reject the would-rather-have-lived-then-than-now option by a ratio of better than two to one.

Explaining the Paradox: The Role of the Subjective

How can this paradox be explained? One can only approach the issue on the basis of conjecture and guesswork. But a pretty plausible account can be developed; the key lies in the consideration that satisfaction and happiness are subjective issues that turn on subjective factors.

The desired account can be given in something like the following terms: an individual's assessment of his happiness is a matter of his personal and idiosyncratic perception of the extent to which the conditions and circumstances of his life meet his needs and aspirations. And here we enter the area of "felt sufficiency" and "felt insufficiency." A person can quite meaningfully say, "I realize full well that, by prevailing standards, I have no good reason to be happy and satisfied with my existing circumstances, but all the same I am perfectly happy and quite contented." Or, on the other hand, he may conceivably (and perhaps more plausibly) say, "I know full well that I have every reason for being happy, but all the same I am extremely discontented and disatisfied." We are dealing with strictly personal evaluations.

In this context one is carried back to the old proportion from the school of Epicurus in antiquity.[8]

$$\text{degree of satisfaction} = \frac{\text{attainment}}{\text{expectation}}$$

The man whose personal vision of happiness calls for yachts and polo ponies will be a malcontent in circumstances many of us would regard as idyllic. He who asks but little may be blissful in humble circumstances. It is all a matter of how high one reaches in one's expectations and aspirations.[9]

This issue of expectations deserves a closer look. People's expectations tend to be geared to the record of their past experience. When improvements are subject to the limits of finitude which generally prevail in human affairs, the situation shown in figure 1.2 results.

Figure 1.2: Expectations of Future Improvement

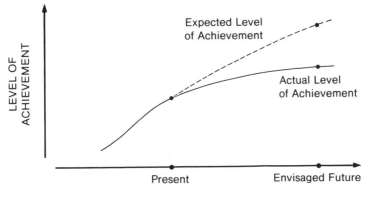

The phenomenon of deceleration is obvious here. We have the usual configuration of an S-shaped, signoid curve of development. When we extrapolate past experience, we see that the result is inevitably such that our expectations outstrip our attainments. The inescapable result is one of dissatisfaction. Things may get "better," objectively speaking, but they don't get better fast enough to meet our subjective expectations.

On this basis, it becomes possible to provide a readily intelligible account for the apparently startling phenomenon of increasing discontent in the present era of improving personal prosperity and increasing public care for private welfare. What we are facing is an escalation of expectations, a raising of the levels of expectations with correspondingly increased aspirations in the demands that people make upon the circumstances and conditions of their lives. With respect to the requisites of happiness, we are in the midst of a revolution of rising expectations, a revolution illustrated by the growth of narcissism and the cult of self-advancement and self-advantage. And, as our Epicurean proportion shows, when increased expectations outstrip actual attainments, even significantly growing attainments, the result is a net decrease in satisfaction.[10]

The paradox mooted above is readily resolved on the basis of these considerations. In the past people were happier because their achievements lived up to — or exceeded — their expectations. With us, even though our level of achievement is higher (and therefore our demands greater), a lower degree of satisfaction is bound to result because of a greater shortfall from expectations. But nevertheless we will not want to exchange our circumstances for the subjectively happier (but objectively less well off) circumstances of the past. It would seem that Americans have come to require more of life to achieve a given level of happiness. Their view seems to be: "To be sure, given what little people asked of life in those 'simpler' days, what they had was quite sufficient to render them happy, or at any rate happier than we are today, we who have more than they. But of course we, with our present expectations, would not be very happy in their shoes."[11]

The Sources of Discontent

Other people can tell a person if he is healthy or in good financial shape more reliably than that person himself. His physician may well be better informed than he on the former score, his tax consultant on the latter. But no one else can tell more accurately whether or not a person is happy. On happiness and its ramifications every man is his own prime authority. Such self-appraisals of happiness are very useful barometers, and the sociologists who design questionnaires have often paid attention to this issue. This tends to produce rather interesting findings, particularly as regards the sources of our discontents, or at any rate their perceived sources. Consider, for example, the data in table 1.2. It emerges that the prime factors that separate the happy from the unhappy are health, aging, and money (which correlates closely with these, given the decline of economic mobility that comes with old age).

Table 1.2: Happiness and Worriment

| Self-classification | | Topics That Respondents Worry "Often" About (% of replies) | | | | | |
	Health	Grow-ing Old	Money	Get-ting Ahead	Work	Marriage	Children
Very happy	16%	5%	39%	35%	48%	12%	38%
Pretty happy	22	8	47	34	51	9	36
Not too happy	42	28	58	37	54	14	31

Source: Figures are derived from data given in N. M. Bradburn and D. Caplovitz, *Reports on Happiness* (Chicago, 1965), p. 55, table 2.27.

Indeed, table 1.3 shows that aging is an especially prominent consideration.

Table 1.3: Self-appraisals of Happiness

| Age | Happiness Categories (% of replies) | | |
	Very Happy	Pretty Happy	Not Too Happy
Under 30	30%	58%	11%
30–39	24	66	10
40–49	25	62	13
50–59	23	59	18
60–69	21	54	24
70 and over	18	52	30

Source: Bradburn and Caplovitz, *Reports on Happiness,* p. 9, table 2.1

Another important factor bearing on happiness has to do with comparisons with others, about keeping up with the Joneses next door.[12] The data of figure 1.3 are interesting in this connection.

Again, a consideration of work satisfaction is illuminating. The accompanying table shows the percentage of people in occupational groups who would choose similar work again.[13]

Professional and white-collar occupations	Percent of Willing Repeaters	Working-class occupations	Percent of Willing Repeaters
Urban university professors	93	Skilled printers	52
Mathematicians	91	Paper workers	42
Physicists	89	Skilled autoworkers	41
Biologists	89	Skilled steelworkers	41
Chemists	86	Textile workers	31
Lawyers	83	Blue-collar workers, cross section	24
Journalists (Washington correspondents)	82	Unskilled steelworkers	21
Church university professors	77	Unskilled autoworkers	16
White-collar workers, cross section	43		

Only professionals in the more prestigious socioeconomic occupations seem comparatively content with their allotted share of the world's work. It is particularly noteworthy that 55 percent of all workers feel that they do less well than persons in other occupations. One is reminded of the statistic that some 80 percent of all motorists regard themselves as better than average drivers.

The empirical facts do not bolster any optimistic view that scientific and technological progress is about to usher in the millennium. The main sources of unhappiness for people seem to be factors like aging and decline in health and vigor, doing less well than one would like, problems in human relationships (particularly in regard to family and chidren), having less prestige than one would like.

Figure 1.3: Workers' Growing Feeling of Inequity

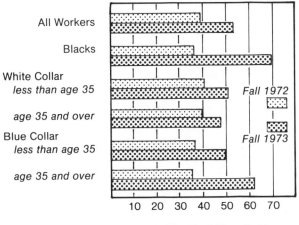

Percent Who Feel They Get Less
Than They Deserve Compared to
Persons in Other Occupations

Source: University of Michigan Institute for Social Research: ISR Newsletter 2 (Summer 1974), 2.

It is clear that factors like these do not readily lend themselves to manipulation by science and technology.

The Influence of the Life Cycle

It is illuminating to correlate people's self-appraisal of satisfaction/dissatisfaction with their age and family status, as shown in figure 1.4. Such a tabulation brings clearly to light where the high-risk factors with regard to happiness attainment are. The advice to offer would seemingly be: be female, be young (under 29), be married, be childless.

These considerations show that much of human contentment/

Figure 1.4: Life Satisfaction of Women and Men at Stages of the Life Cycle (1971)

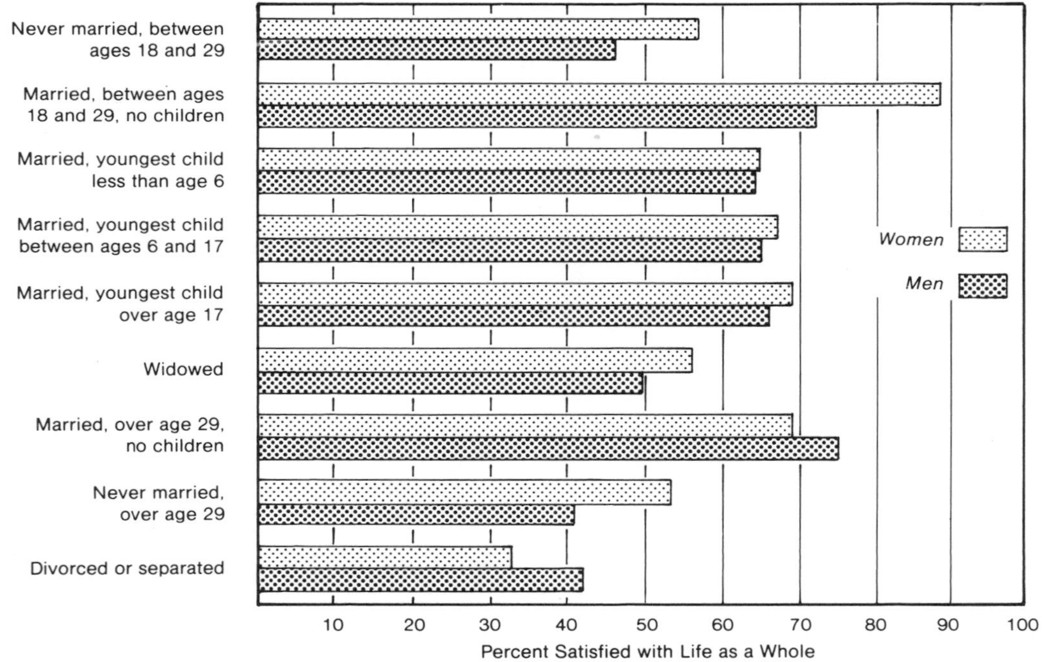

Source: Campbell, Converse, and Rodgers, *The Quality of American Life.*

satisfaction/happiness lies at a level so deep that technological progress can, by comparison, scratch only the surface of life. The capacity of technical progress to contribute to our unhappiness (pollution, overcrowding, system breakdown) is thus much greater than its potential for contributing to our happiness, which seems to turn in large degree on factors like age and human (especially familial) relationships and social interactions that lie largely or wholly outside the manipulative range of science and technology.

Conclusion

A review of some of the main points in our deliberations is now in order:

1. Thinking about happiness and technological progress is apt to be distorted by a deep-rooted tendency to think well of the past. (Every language has a phrase for it: the good old days, *les bons vieux temps,* or *cualquiera tiempo passado fue mejor;* the Romans spoke of *laudatores temporis acti,* "glorifiers of the past.") This is simply an amusing fact that tends to color our thinking about this issue — one that we have to learn to discount for if we are going to think realistically.

2. Contentment and satisfaction seem to depend on very basic elements of the human condition, factors which our technical progress leaves largely untouched and which actually do not admit of ready manipulation.

3. There is what might be called the Fundamental Paradox of Progress: progress produces dissatisfaction because it inflates expectations faster than it can actually meet them. And this is virtually inevitable because the faster the expectations actually *are* met, the faster they escalate.

At this stage it becomes necessary to dwell on the dangers of turning against reason. Science and technology cannot deliver on the $64,000 question of human satisfaction and happiness because, in the final analysis, they simply do not furnish the stuff of which real happiness is made. And here lies the slippery slope of the danger-ous descent from antiscientism to antiintellectualism to irrationalism.

Only reason and intelligence can solve our problems: if we turn against them, we are lost.

And many people seem prepared to turn away from reason and rationality. To quote from one popular book, significantly entitled *Science is a Sacred Cow* by Anthony Standen: "Modern life in this country is highly unnatural. Machines, telephones, radios, vitamin pills, subways, cars, trains, airplanes, elevators, injections, television . . . all products of science, and all intended individually to help us, collectively harry us day and night and drive us to stomach ulcers or the psychiatric ward."[14] In the 1920s and 1930s thoughtful and socially concerned people looked on science and technology as man's best hope and friend. Exactly the same sort of people would now unhesitatingly dismiss this view as hopelessly naive. Indeed science and technology are often seen as "the enemy" of all that is good and humane. Even so informed a thinker as the distinguished bacteriologist René Dubos, in his book *Reason Awake* draws the contrast between the past, when man was threatened by natural forces he could not control, and the present, when our most potent fears are engendered by the malign effects (or side effects) of science and technology.[15]

Surely, great dangers loom ahead along this road. Science, technology, and education in general present the nation with an enormous bill for human and material resources. As long as people maintain the illusion that they afford a royal road to human contentment, they will foot this bill willingly. But what if such disillusionment reaches serious proportions not just with respect to science, but the whole area of the life of the mind?

Science and technology will not, cannot, produce the millennium. And yet in a crowded world of very limited resources we cannot create an adequate setting for human life without them. To recognize that scientific rationality is not a *sufficient* condition for human happiness is one thing—and represents a true insight. However, to reject it as not being a *necessary* condition of human welfare would be a gross mistake. It makes no sense to join the cult of antireason in turning our backs on science and technology. The poor workman always blames his tools; but in this context the difficulties lie not in the tools but in our capacity to make intelligent use of them.

Also, and perhaps most seriously, it is worth dwelling on the dangers of an inflation of expectations. Throughout the history of this country, each generation has addressed itself to life on the premise that the conditions and circumstances of its children would be better than its own. Our faith in "progress" runs deep. What is life going to be like when this expectation is abandoned, or even reversed? Turning expectations around in a zero-growth world is no easy matter. It will be very difficult to get people who have been taught that every day in every way things are getting better and better to accept the idea that the millennium is not around the corner. There is no need to elaborate upon the whole collage of grumbling, discontent, search for scapegoats, political extremism, and so on, that lies in this direction.

Now if the recent escalation of expectations in regard to the requisites of happiness were to continue unabated, then a tragic time of reckoning will come. But man is a creature that learns by experience, and a harsh curriculum of unpleasantly monitory experiences lies ahead.

Finally, let us look back to the initial question: Does scientific and technological progress promote human happiness? I am afraid I have to say no. I do so not because I am a humanistic curmudgeon, but because of the hard facts we encounter when we go into the field and look at the reactions we get from people themselves.

Something akin to a principle of the conservation of negativity seems to be operative in human affairs. It is a cruel fact of life that the achievement of real progress need not be accompanied by any commensurate satisfaction. And there is nothing perverse about this; it is all very natural. Man (as we know him in the West) tends to be a creature of discontents, be they divine or otherwise. The imminent goal once achieved, he simply raises his level of expectation and presses onward to the next goal under the goad of renewed discontent.

One result of this tendency — a result that may properly be viewed as unfortunate — is what might be characterized as the phenomenon of *hedonic discounting*. This is best explained by an analogy. It is a familiar commonplace that the stock market primarily responds not to present economic facts but to anticipations of the future. Making present allowance for foreseeable future economic improvements (or declines), the market has already discounted them by anticipation when they become a reality and so underreacts to or even ignores major achievements when they occur. A parallel phenomenon operates in the context of foreseeable improvements in the conditions of human life: a similar undervaluation of realized achievements in the light of prior expectations. Having expected as much (or generally more), we simply refuse to value very real achievements at their own true worth. Once progress is achieved, it becomes discounted as regards its real contribution to happiness; we have already "raised our sights" in anticipation of its successors. The considerations of this discussion point to the ironic conclusion that advances have in the past, through their promotion of an escalation of expectations, been self-defeating from the standpoint of promoting happiness. The progress that has been made, real though it is, has nevertheless tended to bring in its wake a diminution rather than an increase in the general happiness of people.

Humanists are (or should be) lovers of reason, and one's every rational bone cries out that, ideally speaking, people ought to be happier as their conditions of life improve. But recalcitrant circumstances of the real world indicate that they do not in fact become so. And seemingly for a deep-seated reason. As concerns happiness, progress sets a self-defeating cycle into action:

Improvement ⟶ Aroused Expectations ⟶ Disappointment

It seems that we must bring ourselves to realize, more in sorrow than anger, that it is a forlorn hope to expect technological progress to make a major contribution to human happiness, taken in its positive aspect.

The Environmental Crisis and the Quality of Life 2

We generally think of the environmental crisis as resulting from "too much"—too much pollution, wastage, pesticide, and so forth. But the economist views the problem as one of scarcity—too little clean air, pure water, recreationally usable terrain, agriculturally exploitable land. The answer traditional among welfare economists to problems of scarcity is based single-mindedly on the leading idea of production. But, alas, the things that the environmental crisis leaves in short supply—fresh air, clean rivers, unpolluted beaches, fertile valleys—are not things to which the standard, traditional concept of the production of goods and services applies. The project of producing another planet to live on after we have used this one up is unfortunately infeasible.

Most discussions of the environmental crisis are simply exercises in motivation and social uplift. The lesson is driven home that if we are good and behave ourselves, everything will come out just fine. To adopt more stringent statutes of regulation, to subject grasping enterprise to social pressure, to accept better social values and attitudes, to espouse the program and ideology of planned parenthood or feminism, to hand the control of affairs over to those who are younger and purer of heart—so runs the gamut of remedies which their respective advocates would have us adopt and which, once adopted (so we are told) will put everything to rights. Throughout this stance there runs the fundamentally activistic optimism of the American experience: virtue will be rewarded, and at the end of the sixth reel, the good guys will ride off into the glorious sunset.

The present discussion proposes to dash some cold water on all this. It offers the deeply pessimistic suggestion that, crudely speaking, the environment has had it and that we simply cannot return to "the good old days" of environmental purity. We all know of Thomas

More's futile laments over the demise of the feudal order and the ruralistic yearnings voiced by the romantics in the early days of the Industrial Revolution. Historical retrospect may well cast the present spate of hand-wringing over environmental deterioration as an essentially analogous — right-minded but utterly futile — penchant for the easier, simpler ways of bygone days.

Even to think of the problem as an environmental *crisis* is tendentious. Crises are by definition transitory phenomena: they point toward a moment of decision for life or death, not toward a stable condition of things. The very terminology indicates an unwillingness to face the prospect of serious environmental degradation as a permanent reality, an ongoing fact of life to be contended with in an ongoing way. To take this view goes deep against the grain. It involves a circumstance that is unattractive, and unwelcome. But perhaps it could be granted — at least for the sake of discussion — that the view might possibly be correct and so deserves consideration. Granting this hypothesis, let us explore its implications.

First let us be clearer about the hypothesis itself. It is certainly not maintained that environmental activism is futile, that man cannot by dint of energy and effort manage to clean up this or that environmental mess. Rather, what is at issue is the prospect that we may simply be unable to solve the environmental crisis as a whole, that once this or that form of noxiousness is expelled from one door some other equally bad version comes in by another. My hypothesis, in short, is that the environmental crisis may well be incurable. It just may be something that we cannot solve but have to learn to live with and make the best of.

This hypothesis is surely not altogether unrealistic and fanciful. Basically, the environmental mess is a product of the conspiracy of three forces: high population densities, high levels of personal consumption, and a messy technology of production. Can one even realistically expect that any of these can really be eliminated? Not the population crunch surely. As one character remarked in a "Peanuts" cartoon: "Everybody says there are too many of us, but nobody wants to leave." Nor do things look rosy as regards consumption. People everywhere in the world are clamoring for affluence and a place on the high-consumption bandwagon, and pitifully few are jumping off. Nor, save in certain limited, highly publicized areas, are we likely to get an environmentally benign technology of production and consumption. (Nobody is enough of a bookworm to start eating the evening newspaper, no matter how palatable chemical ingenuity may be able to make it). All in all, it takes much doing to persuade oneself that the day of environmental pleasantness is somehow just around the corner.

Thus the hypothesis in view is not altogether visionary.

Some Ideological Victims

If the continued unfolding of such an environmental crisis occurs, various conceptions integral to the Americal social ideology will have to go by the board, in particular the conceptions of material progress, of technological omnipotence, and of millenial orientation.

Material Progress.   Americans naturally tend to be antipathetic to the idea of a golden age, a time in the past when conditions of life were superior to those of our own day. The concept of progress is deeply, almost irremovably, impressed on the American consciousness. And this is so not just in the remote past but very much in our own day. Take just the most recent period since World War II. Consider the marked signs of progress: (1) the increase of life expectancy (at birth) from sixty-three years in 1940 to seventy years in 1965; (2) the rise of per capita personal income from $1,810 in 1950 to $2,542 in 1965 (in constant [1958] dollars); (3) the increase in education represented by a rise in school enrollments from 44 percent of the five-to-thirty-four-year-old group in 1950 to 60 percent in 1965; (4) the growth of social welfare expenditures from $88 per capita in 1945 to $360 per capita in 1965 (in constant [1958] dollars).

Taken together, such statistics bring into focus the steady and significant improvement in the provisions for individual comfort and social welfare that has taken place in the United States since World War II. The very concept of "the good old days" is anathema to us when the terms of reference are material rather than moral. (Americans have usually been prepared to grant that the Founding Fathers were better—and perhaps even happier—men than their own contemporaries, but they have never been prepared to concede that they were better off.) We have in general had a fervent commitment to the concept of material progress, that everything is getting bigger and better (or almost everything—with such occasional exceptions as trains and restaurant service).

Relinquishing the concept of progress will not come easy to us. It is going to take a lot of doing to accustom us to the idea that the quality of life in this nation is on balance going to get worse or at any rate will not get better. We have had little preparatory background for accepting the realization that in some key aspects in the quality of life the best days may be behind us. It is doubtful, to say the least, that we are going to take kindly to the idea; there may be vast social and political repercussions in terms of personal frustration and social unrest. The British have made a pretty good show of having to haul down the flag of empire. But one may well be skeptical about our ability to show equally good grace when the times comes to run down our banner emblazoned with Standard of Living.

Technological Omnipotence.   A belief in the omnipotence of technology runs deep in the American character. We incline to the idea that, as a people, we can do anything we set our mind to. In a frontier nation there was little need to acknowledge limits of any sort. The realization of finite resources, the recognition of opportunity costs, the need for making choices in the allocation of effort, and the inescapable prospect of unpleasant consequences of our choices (negative externalities) are newcomers to American thinking. The new economic consciousness, the recognition of the realities of cost-benefit analysis, is so recent that it has not as yet come fully home to us. For the course of our historical experience has not really prepared us for these realities of finiteness and incapacity. We expect government to "handle things"—not only the foreign wars, economic crises, and social disorders of historical experience, but now the environmental crises as well. The idea that our scientific technology and the social technology of our political institutions may be utterly inadequate to the task does not really dawn on us. If and when it finally does, the fur will surely fly. (The unhappy fate of the conservationist, retrenchmental sector of Jimmy Carter's energy program is perhaps the first indication of this.)

Millennial Hankerings. Americans have manifested more millennial hankerings than perhaps any other people since the days when apocalyptic thinking was in fashion. The idea that a solution to our problems lies somehow just around the corner is deeply ingrained in our consciousness. Nobody knows the themes to which people resonate better than politicians. And from Woodrow Wilson's Fourteen Points to Franklin Roosevelt's New Deal to the quality-of-life rhetoric of Lyndon Johnson's presidential campaign, the fundamentally millennial nature of our political rhetoric is clear. "Buy our program, accept our policies, and everything in the country will be just about perfect." That is how the politicians talk, and they do so because that is what people yearn to hear. We can accept deprivation now as long as we feel assured that prosperity lies just around the corner. No political campaign is complete without substantial pandering to our millennial yearnings through assurances that if only we put the right set of men in office all our troubles will vanish and we can all live happily ever after. We as a nation have yet to learn the unpleasant lesson that such pie-in-the-sky thinking is a luxury we can no longer afford.

A New Ideology: The Need for Realism

The conception that the quality of life, currently threatened by the environmental crisis, represents simply another one of those conditions of scarcity for which the welfare economists' classic prescription of "producing oneself out of it" applies is profoundly misguided. There is good reason to think that the environmental crisis is not really a crisis at all, but the unavoidable inauguration of a permanent new condition of things. This new condition will engender a reversal of the ongoing escalation of expectations regarding the material conditions of life that is and long has been typical of Americans. In various crucial respects regarding the quality of life we may have to settle for less. This development will unquestionably require considerable ideological revisionism. In particular, it will demand as victims our inclination to progressivism, our Promethean faith in man's technological omnipotence, and our penchant for millennial thinking. What is needed in the face of the environmental crisis at this point, as I see it, may well be not a magisterial confidence that things can be put right, but a large dose of cool realism tempered with stoic resignation. We had better get used to the idea that we may have to scale down our expectations and learn to settle for a lowered standard of living and a diminished quality of life.

The ideological consequences of such a reorientation will clearly be profound. The result cannot but be a radically altered ideology, a wholly new American outlook. What will this be? All too temptingly it may be a leap to the opposite extreme: to hopelessness, despondency, discouragement, and escapism; the sense of impotence and *après nous le déluge*. Such an era of negativism may well be the natural consequence of the presently popular rhetoric of the environmental crisis. And the American people do not have a particularly good record for sensible action in a time of disappointed expectations. Our basic weakness is a rather nonstandard problem of morale: a failure not so much of nerve as of patience.

Yet such a result—despair and disillusionment—is surely unwarranted. It is realism and not hopelessness that provides the proper remedy for overconfidence. Let us by all means carry on the struggle to "save the environment" by all feasible steps. But let us not entertain misguided expectations about the prospects of success, expectations whose probable disappointment cannot but result in despondency, recrimination, and the tempting resort to the dire political measures that are natural to gravely disillusioned people.

The proper stance is surely not one of fatalistic resignation but of carrying on the good fight to save the environment, doing so in fully realistic awareness that we are carrying on a limited war in which an actual victory may well lie beyond our grasp. It has taken an extraordinarily difficult struggle for us to arrive at a limited war perspective in international relations under the inexorable pressure of the political and technological facts of our times. And we have not even begun to move toward the corresponding mentality in the sphere of social problems and domestic difficulties. Yet just this is clearly one of the crucial sociotechnological imperatives of our day.

This conclusion will very likely strike many as a repulsive instance of "gloom and doom" thinking. This would be quite wrong. The moral, rather, is at worst one of gloom without doom. Man is a being of enormous adaptability, resiliency, and power. He has learned to survive under some extremely difficult and unpleasant conditions. By all means, let us do everything we can to save the environment. But if we do not do a very good job of it—and it is doubtful, to say the least, that we will—this is not necessarily the end of the world. We have been in tight corners and unpleasant circumstances before and have managed to cope.

The Allocation of Exotic Medical Lifesaving Therapy 3

Technological progress has in recent years transformed the limits of the possible in medical therapy. However, the very sophistication of modern medical technology has brought the economists' classic problem of scarcity in its wake as an unfortunate side product. The enormously complex equipment and the highly trained teams of experts needed to operate it are scarce resources in relation to potential demand. The administrators of the great medical institutions that preside over these scarce resources thus come to be faced increasingly with the awesome choice: Whose life to save?

A (somewhat hypothetical) paradigm of this problem may be sketched within the following set of definitive assumptions. We suppose that people in some particular medically morbid condition are "mortally afflicted" in that in the absence of treatment it is effectively certain that they will die within a short time. It is assumed, furthermore, that some very complex course of treatment (e.g., a heart transplant) represents a substantial probability of life prolongation for people in this condition. We assume that the availability of human resources, mechanical instrumentalities, and requisite materials (e.g., hearts in the case of a heart transplant) make it possible to give a certain treatment—an "exotic medical lifesaving therapy"—to a number of people that is small relative to those in the mortally afflicted condition at issue. The problem then may be formulated as follows: How is one to select within the pool of afflicted patients the ones to be given the lifesaving treatment in question? Faced with many candidates where effective treatment can be made available to only a few, one confronts the decision of who is to be given a chance at survival and who is, in effect, to be condemned to die.

As has already been implied, the heroic variety of spare-part surgery can pretty well be assimilated to this paradigm. One can foresee the time when heart transplantation, for example, will have become a routine medical procedure, albeit on a very limited basis, since a cardiac surgeon with the technical competence to transplant hearts can operate at most only a few times each week and the elaborate facilities for such operations are bound to exist on a modest scale. Moreover, in spare-part surgery there is always a problem of the availability of the spare parts themselves.

Another example is afforded by long-term hemodialysis, a process by which an artificial kidney machine is used periodically in cases of chronic renal failure to substitute for a nonfunctional kidney in cleaning potential poisons from the blood. There is only a limited number of chronic hemodialysis units, whose complex operation is an expensive proposition. For the present and the near future, "the number of places available for chronic hemodialysis is hopelessly inadequate."[16]

The traditional medical ethos has insulated the physician against facing the very existence of this problem. When swearing the Hippocratic oath, he commits himself to work for the benefit of the sick in "whatsoever house I enter," renouncing the explicit choice of saving certain lives rather than others.[17] Of course, doctors have always had to face such choices on the battlefield or in times of disaster, but there the issue had to be resolved hurriedly, under pressure, and in circumstances which effectively precluded calm deliberation by the decision maker as well as criticism by others. In sharp contrast, however, cases of the type we have postulated in the present discussion arise predictably and represent choices to be made deliberately and "in cold blood."

It should be remarked, to begin with, that this problem is not fundamentally a medical problem. For when there are sufficiently many afflicted candidates—so we may assume—there will also be more than enough for whom the purely medical grounds for lifesaving intervention are decisively strong in any individual case, and just about equally strong throughout the group. But in this circumstance a selection of some afflicted patients rather than others cannot, ex hypothesi, be made on the basis of purely medical considerations. Selection is thus a problem which must somehow be solved by physicians, but that does not make it a medical issue any more than the problem of hospital building is a medical issue. Structurally, it bears a substantial kinship with those issues that revolve about the notorious whom-to-save-on-the-lifeboat and whom-to-throw-to-the-wolves-pursuing-the-sled questions. But whereas those questions are artificial, hypothetical, and farfetched, the issue of lifesaving therapy poses a genuine policy question for the responsible administrators in medical institutions, indeed a question that threatens to become commonplace in the foreseeable future.

A body of rational guidelines for making choices in these literally life-or-death situations is an obvious desideratum. This is an issue in which many interested parties have a substantial stake, including the responsible decision maker who wants to satisfy his conscience (and his potential critics) that he is acting in a reasonable way. Moreover, the family and associates of the patients who are turned away — to say nothing of the patients themselves — have the right to an acceptable explanation. And indeed even the general public wants to know that what is being done is fitting and proper. All these interested parties are entitled to insist that a reasonable code of operating principles provides a defensible rationale for making the life-and-death choices.

1. The Two Types of Criteria

Two distinguishable types of criteria are bound up in the issue of making such lifesaving choices: *criteria of inclusion* and *criteria of comparison*. This distinction requires explanation. We can think of the selection as being made by a two-stage process: (1) the selection from among all possible candidates (by a suitable screening process) of a group to be taken under serious consideration as candidates for therapy; and then (2) the actual singling out, within this group, of the particular individuals to whom therapy is to be given. Thus the first process narrows down the range of comparative choice by eliminating en bloc whole categories of potential candidates. The second process calls for a more refined, case-by-case comparison of those candidates that remain. By means of the first set of criteria one forms a selection group; by means of the second set, an actual selection is made within this group.

2. Essential Features of an Acceptable Selection System

A selection process for the choice of patients to receive lifesaving therapy will be acceptable only when the reasonableness of its component criteria can be established.

To qualify as reasonable, the selection procedure must meet two important structural requirements: it must be simple enough to be readily intelligible, and it must be plausible, that is, patently reasonable in a way that can be apprehended easily and without involving ramified subtleties. Those responsible for lifesaving choices must follow a modus operandi that virtually all the people involved can readily understand to be acceptable (at a reasonable level of generality, at any rate). Appearances are critically important here. It is not enough that the choice be made in a justifiable way; it must be possible for people — plain people — to see (i.e., understand without elaborate teaching or indoctrination) that it is indeed justified, insofar as any mode of procedure can be justified in cases of this sort.

One further requirement is obviously an essential feature of a reasonable selection system: all its component criteria, those of inclusion and those of comparison alike, must be reasonable in the sense of being rationally defensible. Above all, it must be fair — it must treat relevantly like cases alike, leaving no room for influence or favoritism.[18]

3. The Basic Screening Stage: Criteria of Inclusion (and Exclusion)

Three sorts of considerations are prominent among the plausible criteria of inclusion/exclusion at the basic screening stage: the constituency factor, the progress-of-science factor, and the prospect-of-success factor.

The Constituency Factor

Sophisticated therapy is usually available only at a hospital or medical institute or the like. Such institutions generally have clientele boundaries. A veterans' hospital will not concern itself primarily with treating nonveterans; a children's hospital cannot be expected to accommodate the senior citizen; an army hospital can regard college professors as outside its sphere. A medical institution is justified in considering for treatment only persons within its own constituency, provided this constituency is constituted upon a defensible basis. Thus the hemodialysis selection committee in Seattle "agreed to consider only those applicants who were residents of the state of Washington. . . . They justified this stand on the grounds that since the basic research . . . had been done at . . . a state-supported institution — the people whose taxes had paid for the research should be its first beneficiaries."[19]

The Progress-of-Science Factor

The needs of medical research itself can provide a second valid principle of inclusion. The research interests of the medical staff in relation to the specific nature of the cases at issue is a significant consideration. It may be important for the progress of medical science (and thus of potential benefit to many patients in the future) to determine how effective the therapy at issue is with diabetics or persons over sixty or with a negative Rh factor. Considerations of this sort represent another type of legitimate factor in selection for lifesaving therapy. (A borderline case under this head might be a patient who was willing to pay, not in monetary terms, but in offering himself as an experimental subject, say by contracting to return at designated times for a series of tests substantially unrelated to his own health, but yielding data of importance to medical knowledge in general.)

The Prospect-of-Success Factor

It may be that while the therapy at issue is not without some effectiveness in general, it has been established to be highly effective only with patients in certain specific categories (e.g., females under forty of a specific blood type). This difference in effectiveness — in the absolute or in the probability of success — is (we assume) so marked as to constitute virtually a difference in kind rather than in degree. In this case, it would be perfectly legitimate to adopt the practice of making the therapy at issue available only or primarily to persons in this substantial-promise-of-success category. (It is on such grounds that young children and persons over fifty are sometimes ruled out as candidates for hemodialysis.)

To this point, we have held that the three factors of constituency, progress of science, and prospect of success represent legitimate criteria of inclusion for selection for exotic lifesaving therapy. But it remains to examine the considerations which legitimate them. The legitimating factors are in the final analysis practical or pragmatic in nature. From the practical angle it is advantageous, indeed to some extent necessary, that the arrangements governing medical institutions should embody certain constituency principles. It also makes good pragmatic and utilitarian sense that progress-of-science considerations should be operative. And, finally, the practical aspect is reinforced by a whole host of other considerations, including moral ones that countervail against violating the prospect-of-success criterion. Throughout this basic screening stage, then, a legitimating rationale of fundamentally pragmatic considerations is forthcoming.

The appropriateness of each of these three factors is of course conditioned by the ever present element of limited availability. Each is operative only in this context; specifically, prospect of success is a legitimate consideration only because we are dealing with a situation of scarcity.

4. The Final Selection Stage: Criteria of Selection

Five elements must, as we see it, figure prominently among the plausible criteria of selection that will be used in further screening the group constituted after application of the criteria of inclusion: the relative-likelihood-of-success factor, the life-expectancy factor, the family-role factor, the potential-contributions factor, and the services-rendered factor. The first has already been discussed. Let us consider the remaining four.

The Life-Expectancy Factor

Even if the therapy at issue is "successful" in a patient's case he may, considering his age or other aspects of his general medical condition, expect to live only a short time. This matter of the expected duration of future life (and of its quality as well) is obviously a factor that can appropriately be taken into account.

The Family Role Factor

A person's life is a thing of importance not only to himself but to others—friends, associates, neighbors, colleagues. But his (or her) relationship to his immediate family is a thing of unique intimacy and significance. The nature of his relationship to his wife, children, and parents, and the issue of their financial and psychical dependence upon him, are obviously matters that deserve to be given weight in a lifesaving selection process. Other things being anything like equal, the mother of young children must take priority over the middle-aged bachelor.

The Potential-Contributions Factor (Prospective Service)

In choosing to save one life rather than another, the society, through the mediation of the particular medical institution in question—which should certainly look upon itself as a trustee for the social interest—is clearly warranted in considering the likely pattern of future services to be rendered by the patient (adequate recovery assumed), considering his age, talent, training, and past record of performance. In its allocations of complex and expensive therapeutic procedures, society invests a scarce resource in one person as against another and is thus entitled to safeguard its own interests by looking to the probable prospective return on its investment.[20]

It may well be that a thoroughly egalitarian society is reluctant to put someone's social contribution into the scale in situations of the sort at issue. One popular article states that "the most difficult standard would be the candidate's value to society," and goes on to quote someone who said: "You can't just pick a brilliant painter over a laborer. The average citizen would be quickly eliminated."[21] But what if it were not a brilliant painter but a brilliant surgeon or medical researcher that was at issue? One wonders if the author of the obiter dictum that one "can't just pick" would still feel equally sure of his ground, particularly seeing that it is (ex hypothesi) not a matter of *just* picking, but picking for clear and cogent reasons. In any case, the fact that the standard is difficult to apply is certain no reason for not attempting to apply it. The problem of lifesaving selection is inevitably burdened with difficult standards.

The Services-Rendered Factor (Retrospective Service)

One person's services to another person or group have always been taken to constitute a valid basis for a claim upon this person or group, of course a moral and not necessarily a legal claim. Society's obligation to recognize and reward services rendered, an obligation whose discharge is also very possibly conducive to self-interest in the long run, is thus another factor to be taken into account. This should be viewed as a morally necessary correlative of the previously considered factor of prospective service. It would be morally indefensible of society in effect to say: "Never mind about services you rendered yesterday — it is only the services to be rendered tomorrow that will count with us today." We live in very future-oriented times, constantly preoccupied in a distinctly utilitarian way with future satisfactions. And this disinclines us to give much recognition to past services. But parity considerations of the sort just adduced indicate that such recognition should be given on grounds of equity. Moreover a justification for giving weight to services rendered can also be attempted along utilitarian lines. ("The reward of past services rendered spurs people on to greater future efforts and is thus socially advantageous in the long-run future *pour encourager les autres*.")[22]

These four factors fall into three groups: a biomedical factor, a familial factor, and two social factors. With the first, the need for a detailed analysis of medical considerations comes to the fore. The age of the patient, his medical condition, his specific disease, and so forth, will all need to be taken into exact account. These biomedical factors represent technical issues; they call for the physicians' expert judgment and the medical statisticians' hard data. And they are ethically uncontroversial factors. Their legitimacy and appropriateness are evident from the very nature of the case.

Greater problems arise with the familial and social factors. They involve intangibles that are difficult to assess. How is one to develop subcriteria for weighing the relative social contributions of (say) an architect or a librarian or a federal judge? And they involve highly problematic issues. (For example, should good moral character be rated a plus and bad a minus in judging services rendered?) Issues of this sort are strikingly unpleasant for people brought up in times greatly inclined toward maxims of the type Judge Not! and Live and Let Live! All the same, in the situation that concerns us here such distasteful problems must be faced. Unpleasant choices are intrinsic to the problem of lifesaving selection; they are of the very essence of the matter.[23]

But is reference to all these factors indeed inevitable? The justification for taking account of the medical factors is obvious and that of the familial factor is relatively straightforward. But why should the social aspect of services rendered and to be rendered be considered at all? The answer is that they must be taken into account not from the medical but from the ethical point of view.

Despite disagreement on many fundamental issues, moral philosophers of the present day are pretty well agreed that the justification of human actions is to be sought largely (if not exclusively) in the principles of utility and justice.[24] But utility requires reference of services to be rendered and justice calls for a recognition by parity to services that have been rendered. Moral considerations would thus seem to demand recognition of these two factors.[25] To be sure, someone might say that moral considerations ought to play no role in such a context of social polity because they are inherently controversial (philosophers continue to dispute about them — as about all else). But this should not worry us. For one thing, the exclusion of moral considerations is itself a doctrinal stance that is itself no less controversial and disputable. There is no shirking the ethical dimension. We must in this domain, as elsewhere, simply do the best we can in the circumstances at hand.

5. More Than Medical Issues Are Involved

In recent years, an active controversy has sprung up in medical circles over the question of whether nonphysician laymen should be given a role in lifesaving selection (in the specific context of chronic hemodialysis). One physician writes: "I think that the assessment of the candidates should be made by a senior doctor on the [dialysis] unit, but I am sure that it would be helpful to him — both in sharing responsibility and in avoiding personal pressure — if a small unnamed group of people [presumably including laymen] officially made the final decision. I visualize the doctor bringing the data to the group, explaining the points in relation to each case, and obtaining their approval of his order or priority."[26]

Essentially this procedure of a selection committee of laymen has for some years been in use in one of the most publicized chronic dialysis units, that of the Swedish Hospital of Seattle, Washington.[27] Many physicians are apparently reluctant to see the allocation of medical therapy pass out of strictly medical hands. Thus Dr. Ralph Shakman writes: "Who is to implement the selection? In my opinion it must ultimately be the responsibility of the consultants in charge of the renal units.... I can see no reason for delegating this responsibility to lay persons. Surely the latter would be better employed if they could be persuaded to devote their time and energy to raise more and more money for us to spend on our patients."[28] Other physicians strike much the same note. Dr. F. M. Parsons writes: "In an attempt to overcome . . . difficulties in selection some have advocated introducing certain specified lay people into the discussions. Is it wise? I doubt whether a committee of this type can adjudicate as satisfactorily as two medical colleagues, particularly as successful therapy involves close cooperation between doctor and patient."[29] And according to Dr. M. A. Wilson: "The suggestion has been made that lay panels should select individuals for dialysis from among a group who are medically suitable. Though this would relieve the doctor-in-charge of a heavy load of responsibility, it would place the burden on those who have no personal knowledge and have to base their judgments on medical or social reports. I do not believe this would result in better decisions for the group or improve the doctor-patient relationship in individual cases."[30]

But no amount of flag-waving about the doctor's facing up to his responsibility (or prostrations before the idol of the doctor-patient relationship and reluctance to admit laymen into the sacred precincts of the conference chambers of medical consultations) can obscure the essential fact that lifesaving selection is not a strictly medical problem. When there are more than enough places in a facility for lifesaving therapy to accommodate all who need it, then it will clearly be a medical question to decide who does have the need and which among these would successfully respond. But when gross insufficiency of places exists, when there are ten or fifty or one hundred candidates for each place in the program, then it is simply unrealistic to take the view that purely medical criteria can furnish a sufficient basis for selection. The question of selection becomes serious as a phenomenon of scale; as more candidates present themselves, strictly medical factors are increasingly less adequate as a selection criterion precisely because by numerical category-crowding there will be more and more cases whose "status is much the same" so far as purely medical considerations go.

The problem clearly poses issues that transcend the medical sphere because, in the nature of the case, many residual issues remain to be dealt with once all the medical questions have been faced. Because of this there is good reason why laymen as well as physicians should be involved in the selection process. Once the medical issues have been considered, extramedical and specifically social issues remain to be resolved. The instrumentalities of therapy have been created through the social investment of scarce resources, and the interests of the society deserve to play a role in their utilization. As representatives of their social interests, lay opinions should function to complement and supplement medical views once the proper arena of medical considerations is left behind.[31]

One physician has argued against lay representation on selection panels for hemodialysis as follows: "If the doctor advises dialysis and the lay panel refuses, the patient will regard this as a death sentence passed by an anonymous court from which he has no right of appeal."[32] But this drawback is not specific to the use of a lay panel. Rather, it is a feature inherent in every selection procedure, regardless of whether the selection is done by the head doctor of the unit, by a panel of physicians, or some other person or group. No matter who does the selecting among patients recommended for dialysis, the feelings of the patient who has been rejected (and knows it) can be expected to be much the same, provided that he recognizes the actual nature of the choice (and is not deceived by the possibly convenient but ultimately poisonous fiction that because the selection was made by physicians it was made entirely on strictly medical grounds).

In summary, then, the question of selection in the allocation of scarce lifesaving therapy would appear to be one that is in its very nature heavily laden with issues of medical research, practice, and administration. But it is not a question that can be resolved on solely medical grounds. Strictly social issues of justice and utility will invariably arise, questions going outside the medical area in whose resolution medical laymen can and should play a substantial role.

6. The Inherent Imperfection (Nonoptimality) of Any Selection Process

The preceding discussion has argued that four factors must be taken into substantial and explicit account in the operation of a system of selection for scarce lifesaving therapy:

A. *Expectancy of future life.* Is the chance of the treatment's being successful to be rated as high, good, average, or what?[33] And assuming success, how much longer does the patient stand a good chance of living, considering his age and general condition?

B. *Family role.* To what extent does the patient have responsibilities to others in his immediate family?

C. *Social contributions rendered.* Are the patient's past services to his society outstanding, substantial, or average?

D. *Social contributions to be expected.* Considering his age, talents, training, and past record of performance, is there a substantial probability that the patient will — adequate recovery being assumed — render in the future services to his society that can be characterized as outstanding, substantial, or average?

This list has a serious deficiency. It is clearly insufficient for the actual construction of a reasonable selection system, since that would require not only that these factors be taken into account (somehow or other), but, going beyond this, would specify a set of procedures for taking account of them. The specific criteria that would constitute such a system would have to take account of the interrelationship of these factors (e.g., C and D), and to set out exact guidelines as to the weight that is to be given to each of them. This is something our discussion has not as yet considered.

No doubt, there is no such thing as a single rationally superior selection system. The position of affairs in this regard stands somewhat as follows: (1) It is necessary (for reasons already canvassed) to have a system; and (2) to be rationally defensible, this system should take the factors A–D into substantial and explicit account. But (3) the exact manner in which a rationally defensible system takes account of these factors cannot be fixed in any one specific way on the basis of general considerations. Any of the variety of ways that give A–D their due will be acceptable and viable. One cannot hope to find within this range of workable systems some one that is optimal in relation to the alternatives. There is no one system that is "the (uniquely) best"; there are a variety of systems that manage, in one way or another, to accommodate all the various factors.

The situation is structurally very much akin to that of the rules for dividing an estate among the relations of a decedent who leaves no will. It is important that there be such rules. And it is reasonable that spouse, children, parents, siblings, and others be taken account of in these rules. But the question of the exact method of division (for instance that when the decedent has neither living spouse nor living children then 60 percent of his estate is to be divided between parents and 40 percent between siblings rather than dividing 90 percent between parents and 10 percent between siblings) cannot be settled on the basis of any general abstract considerations of reasonableness. Within broad limits, a variety of resolutions are all perfectly acceptable; no one procedure can justifiably be regarded as "the (uniquely) best" because it is superior to all others on the basis of considerations of general principle.[34] Such considerations can only indicate that these factors be taken into account, but they do not determine exactly how this should be done.

7. A Possible Basis for a Reasonable Selection System

Having said that there is no such thing as the optimal selection system, no uniquely correct system of rules for the allocation of livesaving therapy in situations of scarcity, let us now sketch out the broad features of what might plausibly qualify as an acceptable system.

The basis for the system would be a point rating. The scoring would give roughly equal weight to the medical consideration A in comparison with the extramedical considerations B, C, and D, also giving roughly equal weight to these three items. The result of such a scoring procedure would provide the essential starting point of the selection mechanism.

The detailed procedure I would propose—not of course as optimal (for reasons we have seen), but as eminently acceptable—would combine the scoring procedure just discussed with an element of chance. The resulting selection system would function as follows:

1. First the criteria of inclusion described in section 3 would be applied to constitute a first phase selection group which, we shall suppose, is substantially larger than the number n of persons who can actually be accommodated with the therapeutic resources at issue.

2. Next the criteria of selection of section 4 are brought to bear via a scoring procedure of the type described in section 6. On this basis a second phase selection group is constituted which is only slightly larger, say by a third or a half, than the critical number n.

3. If this second phase selection group is relatively homogeneous as regards rating by the scoring procedure, that is, if there are no really major disparities within this group (as would be likely if the initial group was significantly larger than n), then the final selection is made by random selection of n persons from within this group.

This introduction of the element of chance in what could be dramatized as a "lottery of life and death" should be justified.[35] The fact is that such a procedure would bring with it three substantial advantages.

First, as we have argued above (in section 6), any acceptable selection system is inherently nonoptimal. The introduction of the element of chance prevents the results that life-and-death choices are made by the automatic application of an admittedly imperfect selection method.

Second, a recourse to chance would doubtless make matters easier for the rejected patient and those who have a specific interest in him. It would surely be quite hard for them to accept his exclusion by relatively mechanical application of objective criteria in whose implementation subjective judgment is involved. But the circumstances of life have conditioned us to accept the workings of chance and to tolerate the element of luck (good or bad); human life is an inherently contingent process. Nobody, after all, has an absolute claim to therapy in a situation of drastic scarcity, but most of us would feel that we have as good a claim to it as anyone else in substantially similar circumstances. The introduction of the element of chance assures a like handling of like cases over the widest area that seems reasonable.

Third (and perhaps least), such a recourse to random selection does much to relieve the administrators of the selection system of the awesome burden of ultimate and absolute responsibility for life-or-death decisions.

These three considerations would seem to build up a substantial case for introducing the element of chance into the mechanism of the selection system.[36]

It should be recognized that this injection of man-made chance supplements the element of natural chance that is present inevitably and in any case (apart from the role of chance in singling out certain persons as victims for the affliction at issue). As F. M. Parsons has observed: "Any vacancies [in a technologically sophisticated lifesaving program — specifically hemodialysis] will be filled immediately by the first suitable patients, even though their claims for therapy may subsequently prove less than those of other patients refused later."[37] This first-come-first-served aspect introduces an inescapable element of contingency. The realm of the goddess Fortuna has not shrunk since classical antiquity; life is a chancy business, and even the most rational of human arrangements can cover this circumstance over to a very limited extent at best.

Ethical Issues Regarding the Delivery of Health Care 4

Often in life we are "of two minds" about something; conflicting factors pull us in opposite directions and "we can see it both ways," so to speak. We feel ambivalent because there is a good deal to be said on each of two diametrically opposed sides, and the operation of conflicting forces impel us in divergent and incompatible directions. This produces the sort of situation which Marxist theoreticians like to refer to as a *contradiction*. This essay will describe and examine four such conflicts or contradictions in the ethics of health care. These are: (1) the system versus the individual; (2) quality versus equality; (3) the present versus the future; (4) the more narrowly available better versus the more widely available good. Each of these conflicts indicates deep ethical problems that have very practical and pressing implications for policy and procedure.

In seeking to clarify the issues involved here, my aim is to do the job of a philosopher: to raise questions, to sharpen our focus on the issues, and to indicate considerations that must be taken into account. To do this is not necessarily to provide answers or solutions. But it seems nonetheless to be a virtually indispensable preliminary thereto, as well as representing a useful mission in its own right. It is none too soon to examine these problems at the level of general theory before they bewilder us at the level of detailed practice. It is well to clarify the theoretical issues before the push and shove of political controversy blocks any prospect of calm and informative debate.

1. The System versus the Individual

The first tension to be considered relates to the allocation of responsibility for health care and centers on the question, is it to be 45 the individual himself who is primarily and predominantly responsible for seeing that his health-care services are adequately provided for in "the system" that society might institute through design or inadvertence? This question of responsibility clearly poses paradigmatically ethical issues.

In its present-day setting, this problem cannot be handled in the abstract on the basis of general principle alone. We must begin with a closer look at the real-world situation regarding threats to health and the delivery of health-care services. What is striking to anyone who examines the statistics of the causes of death in the United States is the massive role played by conditions traceable to what may simply and honestly be characterized as bad personal habits. Consider the state of affairs portrayed below:[38]

Death Rates for Selected Causes (1972)
(per 100,000 population)

1. Major cardiovascular diseases	494
2. Malignant neoplasms	167
3. Influenza and pneumonia	29
4. Diabetes mellitus	19
5. Cirrhosis of the liver	16
6. Certain diseases of early infancy	16
7. Bronchitis, emphysema, and asthma	16
8. Motor vehicle accidents	27
9. Other accidents	27
10. Suicide	12
11. Violent crime	9

The historical statistics indicate a tendency for death rates to increase markedly from conditions traceable to the impact over many years of certain personal habits: smoking, drinking (and drugs), inactivity, and poor diet. Smoking contributes a substantial fraction to items 1 and 2. Drinking accounts for some of item 1, all of item 5, roughly two-thirds of item 8, and a certain amount of item 9. As to items 10 and 11, it is reliably estimated that one-half of all homicides and one-third of all suicides in the United States are related to the use of alcohol. Inactivity and poor diet contribute massively to item 1.

We all know the propaganda about these so-called diseases of affluence. Middle-aged men who are 20 percent above normal weight run two to three times the risk of fatal heart attack. About 75 percent of lung cancer is caused by smoking. Death rates for heart attacks in men range from 50 to 200 percent higher among cigarette smokers than among nonsmokers, depending on age and the amount smoked. And it would be easy to prolong this statistical litany.

As one considers the causal situation regarding the big killers that represent the major threats to health in present-day America, the realization is forcibly brought home to us that an increasingly improminent role is played by those factors (communicable diseases and nonautomotive accidents) over which the individual has relatively little control and an increasingly prominent role is played by those factors over which the individual in fact has substantial control. Like it or not, we are driven to the conclusion that in very large measure it is the individual himself who alone can and should take the responsibility for delivering the great bulk of health-care services he stands in need of.

This line of thought indicates the unrealism of some of the current discussions on health-care delivery. All too often the issues are posed in the format of a Victorian melodrama. We have the villain (the existing institutional system), the hapless heroine (the poor person being denied medical care), the hero (the state that forces the villain's hand in doing the heroine justice), the appreciative bystander (the cheering public). There hovers before us a monitory example of the poor man with appendicitis being turned away from a mercenary hospital to perish miserably in the streets. This sort of picture cannot withstand critical scrutiny. Even a cursory look at the fact of the situation is enough to show up its unrealism as a pattern for the problems of distributive justice in the medical area.

If such a stress on self-originating initiatives is anything like correct, it bears some very uncomfortable implications. Americans are addicted to, and are pretty competent at handling, issues that require technological solutions and systems-design problem-solving.

We are fascinated by problems that admit of economicopolitical solutions and call for enhanced governmental funding and administrative rationalization. We focus our attention on getting an ambulance service better able to care for heart-attack victims, or on the really big issues of distributive imbalance — geographical imbalance (getting more doctors into rural areas) and service imbalance (getting more doctors into family practice and other primary-care activities). And there is no question that such problems are important and difficult. But important though they are, they yet remain second-order — or even fourth-order — considerations in the larger health-care picture. We must face the sobering but inescapable fact that no foreseeable improvement in medical practice or in the distribution of medical services could make an impact on the health of Americans that would amount to more than a minute fraction of the improvement that could be wrought by the cultivation of better personal habits.

The major issues, it would appear, lie in the sphere of the individual rather than of "the system." And just this is the source of the discomfort. For the real issues involve sensible attitudes and intelligent actions on the level of the individual. They root in the facts of human heedlessness and stupidity. It is clear that such factors are comparatively intractable and do not lend themselves to monetary or administrative resolution. The basic problem is the relative effectiveness of public action versus private values in grappling with the issues. It would be a grave — even if comforting — illusion that a combination of greater governmental outlays and statutory manipulation will prove of substantial avail here. For what we most urgently need is not socialized medicine, or an ampler infusion of public funds into the medical area, or a better distributive system of medical services, or lower hospital costs, but a larger dose of old-fashioned morality or even plain common sense. Unfortunately, the latter items are much harder to come by than the former.

2. Quality versus Equality

In the modus operandi of a system for providing health-care services the fundamental tension between quality and equality crops up from many points of view. One faces in various ways the choice between equal care versus a superior system for providing care in general. It is a question of a smaller pie divided equally versus a larger pie where everyone is better off but the divisions are unequal:

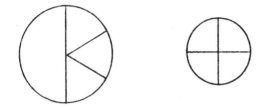

Consider a very concrete example of this. Many of us have admired the National Health Service of Great Britain as an intelligent and efficient component of a combination of public and private medicine, a two-track system of publically and privately financed medical care, both functioning at what has in the past seemed to be a very high level. But some parts of this combination reflected a highly pragmatic compromise made when Parliament instituted the service following World War II: the allocation of some private beds in public hospitals, and the provision of some supporting services (X-ray facilities, laboratory facilities, etc.) by the public to the private sector, in both cases at charges that generally met the actual expenses involved. This pragmatic compromise in fact proved advantageous for public-sector medicine. By exacting a partial subsidy from the private sector it tended to make the public system superior to what it would otherwise have been. And by helping to make medical practice more attractive it has helped to recruit and retain practitioners for British medicine.

On the other hand, there is no question that the two-tier system produces inequities. The private patient has a wider choice of physicians and services than the public health patients do. Moreover, the private patient does not have to wait for nonemergency surgery, while a public health patient may face a delay of many months because of long waiting lists. It is unnecessary to dwell at length on how distasteful this situation seems to the academic ideologues of the political left and their rank-and-file adherents (at any rate in those trades whose unions have not managed long ago to persuade management to provide for private medical insurance). So there has sprung up a powerful and largely successful movement — with nurses and technicians pushed into its foremost ranks — to undo the compromise of the 1940s and create a total divorce between public and private medicine. As could readily be foreseen, purblind pursuit of rigid equality in this context has worked to diminish the capacity of the National Health Service to maintain the quality of service it could otherwise deliver. But this consideration cuts no ice with the doctrinaire devotees of the left.

The preceding characterization has perhaps drawn the lines of this case too harshly for the sake of an example. But, so drawn, it does at any rate illustrate clearly the fundamental point at issue. We face here what is a rather general issue of distributive justice throughout many contexts, namely the issue of *more* versus *more equal.* In certain situations, what is unfair may nevertheless not be unjust in a deeper sense, provided that the unfairness can be shown to work itself out to the general advantage. We face the deep question of whether equality is not bought at too high a price if it can only be had by compromising the qualities of the services available to everybody. The tension in the delivery of medical services between the systemic quality of service and the equality of access in its delivery seems to present a vivid illustration of such a situation.

3. The Present versus the Future

Let us now turn to the third example of a tension or "contradiction" in the health-care area, that between the demands of the present and those of the future — the claims of the living versus those of the yet unborn. The issue can be posed in the form of a resource-allocation question. Specifically, it is the problem of the costs and the benefits of medical research.

Historically, this was a nonissue. Even a single generation ago, the amount of money and talent invested in medical research was a trivial quantity. For example, less money was spent on polio research in the entire prewar generation 1915–1945 than in any pair of years during the decade after 1948. Throughout the early 1960s substantially more was spent on medical research than on nursing-home care throughout the United States. In the early 1970s we spend annually on medical research an amount standing at some 6 percent of the total costs of hospital care, and at more than 12 percent of the sum total of physicians' services. In recent years the increase in medical research and development expenditure has been particularly dramatic, to the point where these costs are a hefty share of the overall pie of total health costs. They have increased from a 3 percent slice of a $26 billion pie in 1960 to a 4 percent slice of a $94 billion pie in 1973—an impressive quadrupling from $.85 billion in 1960 to $3.5 billion in 1973.[39]

A very general and fundamental point about the development of science is forcibly illustrated by biomedical research rather than violated by it. In the course of scientific progress one solves the relatively easy and straightforward problems first and puts the relatively more complex and intractable ones off to the future. As time goes on, problems become more difficult.

Medicine affords an interesting illustration of the extent to which latter-day problems tend to be more intractable than earlier ones and the demand for their solution a vastly greater resource investment. The historical predominance of the diseases that represent the major killers are set out in table 4.1.

Table 4.1: The Killer Diseases

	Percentage of Deaths		
	1900	1930	1960
Heart disease and stroke	15.4	35.9	54.0
Influenza and pneumonia	11.8	9.1	3.9
Tuberculosis	11.3	6.3	0.6
Gastritis, duodenitis, enteritis, and colitis	8.3	2.3	0.5
Children's ailments (measles, diphtheria, whooping cough)	3.8	1.1	0.0
Cancer	3.7	8.6	15.6
Typhoid and paratyphoid fever	1.8	0.4	0.0

Source: Economic Costs of Cardiovascular Disease and Cancer, 1962, Public Health Service Publication No. 947-5 (Washington, D.C., 1965), table 1.

The record of success is doubtless impressive: more than half of the big killer-diseases of 1900 have been virtually eliminated as serious threats. But what is significant is the greater intractability of the problems that remain. Finding a cure for TB or gastritis or diphtheria is still small potatoes compared with finding a cure for today's big killers. In 1962, a total of $1,032 x 10^6$ was spent on medical research in the United States, distributed as follows:[40]

$117 x 10^6$ Cardiovascular disease
$128 x 10^6$ Cancer
$787 x 10^6$ Other

The United States was spending more money (and effort) on cancer in 1962 than it was spending on all medical research in 1950 and, arguably, more than had been spent on all medical research in the history of mankind until 1940. Phenomenally, this massive expenditure has more than doubled to $2,277 x 10^6$ by 1972, and America is currently spending on medical research an amount that is somewhat over 4 percent of what private consumers are expending on medical care.[41] The scale of this research effort is truly impressive. Of course, one tackles the easier problems first—that goes without saying. But the striking thing is the extent to which the later problems become more difficult and demand ever increasing levels of effort for their resolution, in the biomedical area exactly as elsewhere in natural science. It is sobering to contemplate the vast efforts and expenditures of present-day drug research when one considers that the basic research that led to the discovery of penicillin was a shoestring operation costing no more than $20,000.

We thus find ourselves entering into the characteristic condition of a classic law of diminishing returns: costs rise, the significance of results diminish on the average, the waiting times for significant results increase dramatically. Many observers have commented on the ironic fact that, as one acute British writer recently put it: "It is precisely during the last two decades—when scientific medicine is alleged to have blossomed and when the quantity of resources allocated to medical care has rapidly increased—that the decline in mortality that has been associated with the industrialization has tapered off to virtually zero."[42]

Indeed some observers see the situation not as a slowdown but as a stoppage. The eminent Australian immunologist Sir Macfarlane Burnett, who received a Nobel prize for medicine in 1960, after surveying work in the biological sciences concludes:

After working for a year on the present book I cannot avoid the conclusion that we have reached the stage in 1971 when little further advance can be expected from laboratory science in the handling of the "intrinsic" types of disability and disease. There will always be possibilities of improvement in detail but I am specially impressed by the fact that since 1957 there has been no new thought on the handling of cancer, of old age, or of auto-immune disease. The only real novelty has been kidney trans-plantation. . . . None of my juniors seems to be worried as I am, that the contribution of laboratory science to medicine has vir-tually come to an end. The biomedical sciences all continue to provide fascinating employment for those active in research, and sometimes enthralling reading for those like me who are no longer at the bench but can still appreciate a fine piece of work. But the detail of an RNA phage's chemical structure, the place of cyclo-stomes in the evolution of immunity or the production of antibody in test-tubes are typical of today's topics in biological research. Almost none of modern basic research in the medical sciences has any direct or indirect bearing on the prevention of disease or on the improvement of the medical care.[43]

The lesson of these considerations for our purposes is contained in two facts. (1) Research is getting so expensive as actually to direct significant resources from delivery of health care. (2) Progress is so slow and the payoff on research such a long-term issue that the benefit of research may never redound to those actually making the investment but only to their historical successors.

Consider the increased waiting times between the initiation of an intensive research effort and its successful issue in decisive preven-tive or curative instrumentalities. For the communicable diseases that were big killers at the turn of the century this was a matter of a few years and some tens of thousands of dollars; for polio it was a matter of a couple of decades and some $25 millions; for cancer the escalation may well reach billions of dollars stretched over a whole century.

The massive increase in research costs and the elongation of the gestation period exemplify the operation of our third tension, the claims of the present versus those of the future. There can be no question that the massive requirement of medical research for the investment of material resources and skilled manpower leads to a substantial curtailment of our capacity to deliver health-care services here and now.

4. The Question of Resources: Settling for Less than the Best

As medicine advances into regions of ever growing complexity and sophistication, the demands on talent, human services, and equipment grow to a point where the economics of therapy become well-nigh unmanageable. The individual patient has for the most part already arrived at this juncture of economic incapacity in more cases than one likes to think of. And the time when third-party insurers will get there is not all that far away. More and more the real question does not concern the technical issue of what we *can* do, but the economic issue of what we can afford to do.

In the past, the biomedical sciences have flourished at a logarithmic growth rate whose ever accelerating increases cannot be projected into the future. Take the cost of delivering medical services in the United States, for example. If these costs continue to rise at today's rate, they would amount to the whole of our gross national product early in the next century. But we are entering an era of zero growth, which means an era of scarcity in many regards. And here questions of priorities arise, of doing this at the expense of that. Issues of a fundamentally ethical and moral character now arise because the very economics of modern high-technology medicine itself poses moral problems of the most acute and difficult variety.

Throughout recent times American medicine has been governed by the simple precept that the patient deserves the best; where the medical needs of people are concerned nothing less than the very best is viewed as minimally acceptable. This reflects an attitude that is unquestionably admirable. But its practical application is highly problematic. Here, as elsewhere, it can transpire that the better is the enemy of the good. The pushing out of the frontiers of capability can weaken the more mundane, but yet no less important, regions closer to home. The situation in contemporary medicine is reminiscent of the late Roman empire—strength at the outer frontiers and a great multitude of weaknesses and problems in the less exotic regions of the heartland.

The economics of the situation put some very difficult and uncomfortable questions on the agenda:

1. Considering the great expense in manpower and resources involved in work at the frontiers, should there be a major redeployment from research to therapy, adopting the idea that a bird in hand is better than a (pretty costly) bird in the bush?

2. Should we redeploy resources from complex and expensive high-technology intervention to lower-technology therapy and— above all—to preventive medicine?

3. Should we abandon the idea that "only the very best is good enough?" And at just what stage are we to shift from an *optimizing* to a *satisficing* rationale in the delivery of health care (to use Herbert Simon's term)?

The present discussion has drawn attention to four conflicting tensions in the area of the delivery of health-care services: (1) the extent to which responsibility can be imputed to "the system" for delivering medical services versus the extent to which it does and should rest with the individual; (2) the relative desirability of enhancing the overall quality versus improving the equality-of-access to medical services; (3) the claims of the present versus those of the future as an issue posed by the escalating resource demands of biomedical research; (4) the related, economically rooted problems posed by the principle that the patient deserves the best and the issue of expensive (and so limited) high technology intervention versus less costly (and so more pervasively available) lower-technology therapy. Each of these cases has the classic form of a moral dilemma In each instance we are torn in a conflict of two discordant desiderata—individual responsibility versus social concern; quality of service versus equality of opportunity; present welfare versus future benefits; deeper effectiveness versus wider effectiveness. All four issues pose problems of a very general structure that arise as ethical issues in many areas. But they seem to me to pose particularly relevant and increasingly pressing problems in the medical context because the stakes are so high: the physical welfare and indeed the very lives of people are at stake.

Although nothing said here can be claimed to provide or even to point toward any manageable solutions to these problems, there nevertheless are some rather obvious morals that emerge more or less spontaneously once the issues are spelled out clearly. (1) Our concern for the delivery of health-care services has been too system-oriented. People as individuals can and should carry a far heavier burden of responsibility here. (2) The desideratum of equality, no matter how legitimate a value in its own right, should not be promoted in so doctrinaire and unthinking a way as to dilute the capacity of the health-care system to serve the community at large. (3) We should acknowledge as real and authentic the claims presently made upon us by the concerns of the future through the mediation of research. (4) Nevertheless, this can and should be done with open eyes, with the realization that these claims may become very large indeed if substantial progress is to be made as one penetrates further into the area of diminishing returns.

Ideals, Justice, and Crime 5

This essay will consider the relative role of theoretical ideals and harsh realities in relation to justice. What sort of a thing is justice or, for that matter, other things of its kind, like liberty or equality? The answer is that they are clearly ideals. Accordingly, let us begin by considering what ideals are, how they work, and what can be expected of them. Thereafter, in the second part of the essay, we shall apply these general considerations about ideals to justice in particular and examine in this light the major obstacles to realizing a greater degree of justice in American society under present-day conditions.

Ideals

The word *ideal* used as a noun, is quite versatile. The dictionary accords a wide spectrum of senses to the term, of which the one relevant for present purposes may be defined as follows:

an aspect of states of affairs that is to be valued for and desired on account of its embodiment of an element of excellence; a noble desideratum relating to the human condition. Examples: liberty, equality, and fraternity; liberty and justice for all; well-roundedness and the full development of human potential; humility or courage or charity.[44]

An ideal in this sense represents an important human value that we would — or at any rate feel we should — like to promote and to see realized more fully, and whose enhancement is viewed as an inherently worthy objective in the management of human affairs.

What do ideals do for us? What role to they play in the scheme of things? The answer lies in the consideration that man is a rational agent. He can act and he must choose among alternative courses of 57 action. This crucial element of choice means that our actions will be guided by ends and aims, purposes and objectives. To some extent, these goals will be fixed by considerations of "necessity" relating to survival and physical well-being. But to some extent they can, and in an advanced condition of human development they must, be guided by necessity-transcending objectives. Man's "higher" aspirations come into play—his yearning for a life that is not only bearable, and not merely secure and pleasant, but is also meaningful in that it has some element of excellence or nobility about it. And this is where ideals come in. They are the guideposts to action aimed at these excellence-oriented objectives in human affairs.

Ideals always involve an element of *idealization*. They are aspects of excellence, taken in isolation as apart from other aspects. When we consider one such desideratum in separation from others, we engage in an act of abstraction—we put all other considerations aside for the time being. This aspect of idealization attaches to all ideals. It has important consequences. It means that we do not encounter ideal situations in actual experience. Ideals are not merely "things of the mind," but visions of things about which there is always an element of the visionary. This links them to idealism in the other sense of this term—not that of an attachment to ideals but that of the doctrine of the fundamentality of mind-made artifacts. They are incapable of "genuine fulfillment." Circumstances as we actually encounter them do not—nay cannot—exhibit the complete and unqualified realization of ideals.

An ideal tells us what is best in some respect or other. When we know the ideal state, we have before us what is the very best of possible alternatives. But this is not enough. For to apply any such principle in practice we must know which of several feasible nonideal alternatives is to be preferred; we must know not only what is the best, but must be able to determine—in the great majority of cases, at any rate—which of several alternative possibilities is "the better." A principle of evaluation is not adequate if it merely depicts a theoretical ideal that we cannot apply in practice to determine which of several putative possibilities comes "closest to the ideal." (How far has the beginner come toward learning how to evaluate bridge hands when he is told that the ideal holding consists of the four aces, the four kings, the four queens, and a jack?)

By themselves, ideals are thus a very incomplete guide to action. One must have the practical sense needed to effect a working compromise among them. It is a point well worth stressing that an ideal morality — a framework of ideals — is in itself far from sufficient to furnish the guidance we need in the proper conduct of our evaluative affairs.

What is needed for making effective use of our ideals is something to which writers on ethics and value theory have been loath to address themselves, a criterion of comparative merit for suboptimal alternatives. To proceed intelligently in the sphere of values, we must be capable not simply of absolute idealization (i.e., of knowing what the ideal is), but also of relative evaluation (i.e., of determining which of the actually available suboptimal alternatives is to be regarded as the most satisfactory).

In thinking about an ideal, one should never forget that other, "competing" ideals are there as well. Ideals, like values and desiderata in general, do not exist in a vacuum. They do not operate in isolation. They can be pursued only in a complex setting where other "complicating" factors are inevitably copresent. Every ideal coexists alongside others within the context provided by what could be characterized as a system of ideals.

Ideals must accordingly be coordinated with one another. Each ideal can — and should — be pursued and cultivated only insofar as this is consonant with the realization of other values. The pluralism of ideals, the fact that each must be taken in context, means that in the pursuit of our ideals we must moderate them to one another. Whenever multiple desiderata coexist they stand in contextual interaction; we cannot appropriately pursue one without reference to the rest.

Safety is a prime desideratum in a motorcar. But it would not do to devise a "perfectly safe" car whose maximum speed is 1.75 mph. Safety, speed, efficiency, operating economy, breakdown avoidance, and so forth, are all prime desiderata of a motorcar. Each counts but none predominates in the sense that the rest should be sacrificed to it. They must all be combined in a good car. The situation with respect to ideals is altogether parallel.

The pluralism and interactionism of ideals means that a narrow focus upon a single ideal is not very helpful in guiding action. Ideals must stand in balance. To stress one to the neglect of the rest is ultimately self-defeating. The pursuit of ideals is profitable only within the setting of a concern for the overall "economy" of the system of ideals, providing for the mutual interaction of its component desiderata in the light of mutual constraints. An ideal must thus be seen to figure as simply one component within a framework of interrelationship which makes it possible to strike a reasonable balance between the different and potentially discordant demands of various concurrent ideals.

Two points must be carefully distinguished in this connection. The first is the simple and essentially economic point that the resources we commit to the pursuit of one ideal cannot be committed to that of another; in a world of limited resources we must thus make choices. This point is true—but also rather trivial. The second is more subtle. It is that the realities of the world are such that ideals stand in systematic interaction of the sort exemplified in the motorcar example. And this means that their realization is inherently limited by the intrinsically interactive nature of things.

Different aspects of "our imperfect world" are at issue: in the one case we have to do with man's limited resources, in the other with man's limited power. Even were our resources unlimited, there would nevertheless remain drastic, nature-imposed limits to the extent to which man's wish and will can be imposed on the realities of the world. The stress on ideals must be tempered by recognition of this inherent feature: they balance off, one against another.

This feature of our ideals (and of our desiderata in general) means that the contrivance of an evaluatively suitable mix of life arrangements is a design problem, akin to that of landscape gardening or interior decoration. One can have too much of a good thing. Simplicity is fine but can lead to barrenness; variety is good but can lead to clutter. Ideals can be similarly overstressed. Justice is a splendid thing; but *fiat justitia, ruat caelum* is not an appealing precept.

This aspect of the inherent interaction-limitedness of ideals has important consequences. It means that ideals can only be pursued within the limits of the possible in a complex and no doubt imperfect world. And so, while ideals can and should be striven after, they can never be realized fully or perfectly or completely because the attempt to achieve such realization would mean an unacceptable sacrifice of other ideals. Their pursuit must be conditioned by a realistic appreciation of the intrinsic limits: there is a point at which further pursuit would produce unacceptable sacrifices elsewhere and thus prove counterproductive in the larger scheme of things.

In the pursuit of ideals, unrealism is thus a constant danger. There is an object lesson to be learned from the case of the man so intent on the cultivation of his pet ideal that he fails to realize that it is not worth having if its achievement blocks the way to other desiderata. (This might be called the "Monkey's Paw" effect after the classic short story by W. W. Jacobs.) Such unrealism is implicit in the somewhat negative note struck in common language by the use of "idealistic" to characterize a person who has an exaggerated (unrealistic, unwarranted) view of the extent to which an ideal can be brought to realization — without producing untoward side effects.

The cultivation of ideals must thus be tempered and conditioned by the recognition that the pursuit can be overdone. Even as we can make the car so safe that we exact an unacceptable reduction of (say) its economy of operation, so we can emphasize public order in the state in such a degree as to compromise individual liberty, for example. Our dedication to ideals must be tempered by the recognition that in their cultivation there comes a juncture, the point of no advantage, where "the better is the enemy of the good."

A very important difference thus exists between a *compromise* of one's ideals and a *betrayal* of them. The former occurs in the cultivation of one's ideals when one tempers or limits their further pursuit because their interaction with other ideals requires some mutual accommodation. To press further with the supposedly "compromised" ideal would produce unacceptable sacrifices elsewhere within the framework of ideal-endorsed objectives. In this regard, the compromise of ideals is inevitable, "realistic," and nowise reprehensible.

The betrayal of an ideal is something very different. It is to sacrifice it for unworthy reasons — greed, convenience, conformity, or the like, to desist in its pursuit in circumstances where pressing on could and would actually further the good cause. Betrayal involves not going as far as one ought; compromise, by contrast, involves not going further than one ought. The two things are clearly very different.

It is important—crucially important—for a person to have ideals. The person without them is like one who wanders through life lost, bereft of any sense of direction. But having a destination, like having ideals, is not enough: it does not resolve the practical choices that confront us on a moment-to-moment basis. Having a destination is not much help; we must know about life's twists and turns as well. If one "doesn't know how to get there from here," then having even an excellent destination is quite pointless.

Justice

Let us now change course and consider the bearing of these abstract considerations about ideals upon one very particular ideal, that of justice in general and criminal justice in particular. For justice is clearly an ideal, and so, in particular, is social justice with its idealized conception of "the just society." (The concept of "liberty and justice for all" is in fact one of the great historic ideals in the evolution of human aspirations.) Accordingly, certain considerations regarding justice at once follow in the light of the preceding analysis of ideals in general.

To begin with, we cannot reasonably expect justice to become an actually realized aspect of the existing state of things. Justice, like any ideal, is a goal to be striven after; it is not a condition into whose full possession we can ever hope to enter. John F. Kennedy put it simply: "Life is unjust." It is counterproductive to espouse millennial expectations, because this leads to a frustration and defeatism that would impede our realization of suboptimal but perfectly feasible objectives. With justice, as with other ideals, it can happen that the better is the enemy of the good.

Another point is equally important. Justice, like any ideal, will have to stand alongside other ideals. And these may countervail against it in such a way as to require the sort of mutual coordination which we have seen to be typical with respect to ideals in general.

Now there is one particular conflict within the ideal of justice itself to which I wish to call special attention. This is the tension between personal justice—justice as it pertains to the rights, claims, and interests of particular individuals—and communal justice—justice as it pertains to the rights, claims, and interests of people in general within what one might call the "community at large." There is a potential conflict between retail justice construed as an ideal for individuals on the one hand (heed for the rights of individuals, the safeguarding of particular interests) and wholesale justice as an ideal for the public at large, pertaining to the desiderata of communal welfare ("the general advantage"; the safety or security of the public at large).

This potential conflict between personal and communal justice can become particularly acute in the context of a legal system of criminal justice. The human condition being what it is, we cannot perfect such a system with respect to every desideratum. If we are overly sensitive to protecting the rights and interests of the accused individuals who are caught up in the system, we enlarge the numbers of potentially dangerous criminals who escape its net to put the community at risk; if we are too zealous about insuring that all of the guilty are caught in the net we must tighten its mesh to a point where it endangers the rights and interests of the guilty and perhaps of some of the innocent as well. This tension is clearly an instance of problems arising from the coordination of ideals. It is a matter of getting the balance right in resolving problems of intrasystemic competition, of getting enough of one ideal without the unfortunate side effect of the insufficiency of another.

There is little difficulty in discerning the shifts that have occurred historically in attempts to achieve a suitable balance between the pull of atomic and of molar demands. The rights and claims of individuals and the interests and general advantage of the public at large have always been a matter of difficult and precarious balance. And over the years, the pendulum has swung back and forth between too great an emphasis in the one direction or the other.

When one considers where the United States stands in the 1970s in this regard, there can be little question that we do not in fact have the balance right, and that we are currently intent—exaggeratedly, unrealistically, and almost foolishly intent—on cultivating the individual-oriented sector of justice at the expense of its socially-oriented sector. On the issue of individual claims (the rights of the accused, for example) in contrast to the interests of the public (the personal safety and security of people), we have come to lean far too much to the side of the individual.

Crime

A clear symptom (and effect) of this imbalance is the social disorder that pervades contemporary America. Its most striking manifestation is crime, and particularly and most strikingly the proliferation of crimes of violence. For a society nominally at peace with itself, we have moved shockingly far along the road toward realizing "the violent community," as shown graphically in table 5.1.

Table 5.1: Crime and Criminality in the United States

	1960		1973		
	No.	Per 1000 Pop.	No.	Per 1000 Pop.	% Increase 1960-73
Criminal offenses known to the police					
Total	3,364,000	19.00	10,192,000	48.00	258
Homicides	8,464	0.05	20,465	0.10	100
Total arrests	3,679,000	20.00	6,500,000	31.00	52
Prisoners at end of year (state & federal)	212,953	1.19	204,211	0.98	-17

Source: Statistical Abstracts of the United States, 1975

These figures bear interesting comparison with public expenditures for law enforcement (in millions of current dollars):

1960		1973		
Amount	Per Capita	Amount	Per Capita	% Increase 1960-1973
$3,349	$18.54	$12,985	$61.71	332%

We may let the cost/effectiveness aspect of these figures pass in charitable silence.

These statistics lead back to our initial question. They indicate that America is in a very fundamental respect a grossly unjust society, a society where the most basic rights of people within the general community—their rights to security in possessions, and indeed even safety in life and limb—are very inadequately safeguarded.

One aspect of this condition of things deserves special emphasis. Ordinarily we tend to think, with a view to the course of recent history in much of the world, that justice is preeminently a matter of preventing the state from infringements upon the legitimate rights and interests of individuals. But the circumstances of the contemporary American context are rather different in this regard. In our present-day situation, by far the gravest threat against the people's legitimate rights and interests comes not from the state but from other individuals—the growth of "big government" and the pervasive influence of the state in our lives notwithstanding. The gravest block to the just society at this stage of American history is not the abuse of public power, but the greed, malice, and folly of our individual fellow men.

Think of the hue and cry that would ensue today if some agency of government were to abridge even a single life without good and sufficient reason, without due process and the benefit of extensive legal proceedings. Think of the vast and elaborate machinery we have created to assure that justice is done—and seen to be done—to the individual at the hands of the state. In matters large and small we have some reasonable security that the state will treat us justly, and we usually have some means of recourse when it does not.

But where our fellow individuals are concerned, the matter is very different. Consider the enormous toll of criminal violence, the thousands killed, maimed, raped, and robbed by the malicious individual initiative of their fellow men. It is no exaggeration to say that the injustices perpetrated upon Americans by all the organs of the state, at every level, pale into virtually trivial insignificance as compared with the toll taken by violent criminality. The simple (though cruel) truth is that no other factor even begins to compare with crime in contributing to the circumstance that present-day America is removed very far indeed from the ideal of a just society. Ironically, exaggerated fears about the threat to individual (personal) justice posed by the "big state" have contributed not a little to this gross abridgement of social (communal) justice.

What can be done to remedy this situation? I certainly cannot claim to have the answer. And there is no reason why I should, being neither a sociologist concerned with uncovering the causes of criminality, nor a criminologist concerned with determining its treatment. The job of the philosopher is to identify problems, to clarify their structure, and to evaluate their bearing. His task is diagnosis; the cure involves manipulative talents outside his sphere. Of course, I do have a few ideas—ideas, to be sure, for which no particular originality can be claimed.

The overall object is, of course, the reduction of violent crime. And this has two aspects: preventive anticipation and ex post facto treatment. The former involves measures of social amelioration (the reduction of poverty, ignorance, and alienation), the latter involves criminal justice per se. Now insofar as the criminal justice system can make its contribution to the desideratum of crime reduction, we would be well advised to rethink the role of the three main parameters here: rehabilitation/reform, deterrence, and incapacitation.

Let us begin with the first of these approaches to the control of criminality, rehabilitation (as it is optimistically called). The painful experience of many decades has finally brought home the fact that we simply do not know how to go about the rehabilitation and reform of criminals. It would be a fine thing if we could do it—but then it would be very nice if money grew on trees or eating apples cured cancer. The cruel fact is that we might as well write rehabilitation off in our thinking. Nobody has any workable ideas of how to go at it on the scale requisite for an effective criminal justice system in the United States.

Deterrence is something else again. It hinges on three major considerations: the likelihood of apprehension; the prospect of conviction (if apprehended); the severity of the penalty (if convicted). When people talk of deterrence, they generally make the great mistake of thinking in terms of punishment alone. But the other two factors are far more crucial, and a great deal could surely be done here. With respect to the prospect of conviction, for example, we might do well to take a leaf out of the book of our English cousins. Their system of criminal justice has for some time adopted the practice of so-called majority verdicts, where the votes of ten out of twelve jurors suffices for conviction. Such an alteration of present practices would diminish the prospect of a successful resort to bribery, intimidation, or of the exploitation of human foibles.

As regards the third factor, incapacitation, a good case can be made for holding that it deserves to be taken more seriously in considering the rationale of punishment than has generally been the case in the past. Muggers, burglars, and rapists cannot practice their trade in prison: each day in jail is a day on which a criminal cannot victimize other citizens in the community at large. Of course we must house him and clothe him and feed him at a cost that may reach $25,000 per man-year. But that may be a tolerable price to pay in modern society for greater security of life, limb, and property. And in any case, we must learn to weigh in the scales of justice not only the rights and claims of the criminal offender, but those of his past and potential victims as well.

Since I am a philosopher, I can ask that others devise the practical steps necessary to grapple with the problem of violent criminality. But it warrants stress that the problem is today of such a magnitude that we must approach it with a more determined and enterprising spirit than we have done in the past and are in course of doing at the present. And in coming to grips with this issue we have to put away our ideological blinders and deal with the world as it is, not as we would like it to be. Above all, we must approach the matter pragmatically and in an empirical spirit. There is no point in seeking to formulate pet solutions on a basis of "general principles." Dogmas and ideologies will not help; what is needed is a pragmatic approach based on intelligent experimentation and a careful heed of the empirical facts.

In particular, we would do well to take advantage of the important opportunity given us by the fact that we have fifty different states with different sets of laws and law-enforcement processes. One often hears this diversity lamented, but it is a blessing in disguise. It is the height of foolishness to insist on uniformity when one has no real assurance regarding what to be uniform about. And we can only learn from the experience of others in circumstances where those others are in a position to have experiences different from our own.

Economics Versus Moral Philosophy: The Pareto Principle as a Case Study

6

On issues of social policy, economists traditionally tend to enroll themselves in the ranks of the philosophical utilitarians. They view questions of social choice in terms of sharing out goods and evils, benefits and sacrifices, in short, as distributions of positive or negative "utility" to the members of the society. And they incline to accept as the determining consideration the traditional utilitarian guidepost of "the greatest good of the greatest number."

On this basis, a pivotal role comes to be played by a principle that goes under the name of the great Italian economist Vilfredo Pareto (1848–1923). Its key components are as follows:

Definition.   One distribution of utility to the members of a society is a "Pareto improvement" upon another if it is such that some fare better and none fare worse.

Definition.   A distribution is "Pareto optimal" within a range of alternatives if it represents Pareto improvement over every other member of this set.

Thesis.   Whenever one alternative represents an overall distribution of utilities to the members of a society that is Pareto optimal within a set of its rivals, then the "socially rational" thing to do is to prefer this alternative over the rest.

There is little that economists of different schools and persuasions agree on almost universally, but the Pareto Principle seems to be among the few exceptions to this rule. Virtually without exception, economists, decision theorists, social-choice theoreticians, and the like, are inclined to espouse it as a well-nigh self-evident truth. It is viewed as so secure in itself as to qualify as a touchstone by which the adequacy of social-choice mechanisms can be assessed. Accordingly, it has become established dogma in the field that, when a Pareto optimal alternative exists, then only a social decision process that leads to this alternative can possibly qualify as "rational" (appropriate, justified, or what have you).

This discussion will use the Pareto Principle as a case study to exhibit the substantial divergence in perspective and approach that separates economists (decision theorists, social-choice theoreticians, etc.) from traditionally oriented moral philosophers. Specifically, it will attempt to do three things: (1) to maintain that the Pareto Principle is by no means a self-evidently valid truth from the standpoint of traditional social and moral philosophy; (2) to argue that there is much to be said for the position of the moral philosophers—that they are not just being obtuse or obscurantist; (3) to show how this disagreement is symptomatic of a deeper difference of approach, a difference based on issues which the economist or decision theorist can only neglect at the price of the acceptability of his findings and the ultimate adequacy of his labors.

The Rationale of the Pareto Principle: How Might It Be Legitimated?

How would an adherent to the Pareto Principle argue for it if actually pressed to do so? Presumably he would proceed by invoking the idea of the collective will of the community as conceived within a democratic ethos. The argumentation would run roughly as follows: "The rational person desires his own advantage, and so the rational society must, by compilation, desire the case-by-case advantage of its constituting members. And this conception of individual advantage is best and most efficiently implemented in the voting process. By use of voting to determine what members of society individually prefer (on the basis of judicious assessment of their own self-interest), we can determine effectively how that society collectively construes its own best advantage."

To all appearances, this approach immediately yields a straightforward validation of the Pareto Principle. For if the choice between two alternatives, one of these being a Pareto improvement over the other, were put to a vote, then, since some gain, others fare the same, and none lose, one would expect there to be some votes in favor, none against, and various indifferent abstainers. And, accordingly, a Pareto optimal alternative seems bound to emerge victorious from such a social decision process based on a rational exploitation of the voting process.

This line of defense is certainly plausible. But is it decisive? Let us examine the voting approach to the preferability assessment of distributions a little more closely. Is it really all that clear that a rationally preferable result must emerge from such a decision process?

The Voting Resolution: A Closer Look at the Tacit Presupposition

To begin with, it is necessary to distinguish carefully between two very different sorts of votes with two very different sorts of ballots.

Case 1: Blindered Self-Interest.   Would you rather have an alternative that gives *you* the return *x* or which gives *you* the return *y,* no information about how others fare being provided?

Case 2: Holistically Construed Communal Interest.   Would you rather have an alternative that gives such-and-such a distribution across the whole population (you yourself getting the return *x*) or one that gives a certain variant distribution (you yourself getting the return *y*)?

With case 1, the only thing at issue is (in effect) an expression of preference with respect to one's own share. With case 2, the situation is very different. Here there is a choice between two all-inclusive distributions—distributions within which, to be sure, one's own case happens to figure. The first case is so set up that "one's own share" becomes the be-all and end-all. But the second case involves a vote not just on shares but on policies, and not just on a distribution as such, but on the principles involved—at least implicitly and by indirection.

These two approaches need by no means yield the same voting outcome. For it is possible and indeed desirable that a person, even a "rational" person, should consider in assessing the preferability of a distribution not just how he himself fares but how others fare in general. And what is at issue here is not just a matter of particularized fellow feeling or of envy, but a disinterested vision of "the good society," a vision that actually and rightly finds expression in much of our social appraisal.

With these ideas in mind, let us explore the linkage between Pareto optimality and social rationality somewhat more closely. It is relatively clear that the distribution to 101 people consisting of a hundred ones plus one,

$$\overbrace{1, 1, \ldots, 1, 1,}^{100 \text{ times}} \tag{1}$$

is improved upon (from the angle of "social preferability") by the distribution

$$\overbrace{1, 1, \ldots, 1, 2.}^{100 \text{ times}} \tag{2}$$

But does this process continue ad indefinitum? We, or the rational man, might also have to prefer to distribution 1 its rival:

$$\overbrace{1, 1, \ldots, 1, 1000.}^{100 \text{ times}} \tag{3}$$

But is it all that straightforwardly clear that (3) improves upon (1)? Does it really improve matters to let wider and wider disparities open up between the lot of the hapless many and the fortunate few (to put the issue in a somewhat prejudicial way)? This is the inexorable consequence if the question were one of voting exclusively on the basis of what we have characterized as blindered self-interest, if one put aside, that is, any concern for how others are faring under the distribution at issue, be it on the basis of moral sentiment like sympathy or envy or on the basis of disinterested principles like equality or justice. But just this supposition is surely untenable.

Given that the justification rests on such a basis, it signally fails to realize its objectives. Voting on the basis of narrowly construed and atomistically separated self-interest may or may not be "realistic" as an account of the empirical realities, but it is neither inevitable in practice nor desirable in theory.

It does not require profound and subtle reasoning to see that an argument premised on blindered self-interest cannot validate the social propriety (or social rationality) of a measure. It is surely both realistic and justified that people should take account of how others fare in assessing a distribution. And in rejecting blindered self-interest as a basis for procedure, their (rational) voting preferences would surely take account not just of what they themselves obtain in atomistic isolation, but the sort of environment (world, society) they get to live in. Distributions would be appraised not just in terms of their implications for "number 1," but in terms of their implications in general; they would and should be evaluated as exemplifications of a social principle or embodiments of a social policy.

A Variant Defense: The Slippery Slope Argument

In the face of such objections, a defender of the Pareto Principle might perhaps take a different line. He might seek to approach the issue via the slippery slope argument: "If you prefer distribution (2) to distribution (1), then a process of 'natural iteration' sets in and you will be constrained by 'transitivity of preference' to prefer distribution (3). Now the Pareto Principle is safely home." But does this argument work? As the argumentation is pressed further and further, one is surely going to reach the point of the riposte: "Enough is enough!" And this stance is not without its justification.

To be sure, it would be a matter of peevishness and regrettable envy not to fail to prefer distribution (2) to distribution (1). But does this concession force through the thick end of a wedge that means we must also prefer distribution (3)? Is one deficient in rationality (or social rationality) in failing to concede this preferability? Surely not. Even if one puts envy aside as unworthy and not deserving to be taken into account, there remains that part of the "sense of justice" which involves at least a modicum of egalitarianism (in some guise or other). Accordingly, it becomes necessary to complicate the picture of the social-preferability assessment of utility distributions by looking at this issue not just by a microscopic survey of how individuals fare, but by a macroscopic regard for the global bearing of the distributions in terms of the impact of general principles.

And when one looks at the issue from this point of view, it looks very different indeed. The force of the slippery slope argument is dissipated by the consideration that one surely can and should ultimately "draw the line," that at some stage considerations of general justice come to overbalance those of individual advantage. The fact that a line is hard to draw—or rather, that its drawing at one precise spot is hard to justify relative to other, nearby ones—does not imply that it should not be drawn at all.[45]

The Orientation of the Traditional Moralists

Our focus on the general principles implicit in various particular distributions highlights the difference between the approach of economists on the one hand and old-school moralists on the other.

The old-school moralists see the focus of considerations of social rationality as aimed not at distributions at all, but at the level of the general criterial principles or policies that can be exemplified or contravened by distributions. They insist upon inserting another level of consideration as intermediary between the assessment of rational preferability and distributions. In their view, the issue of rational social decision making is, in the first instance, not a matter of distribution assessment at all, but of the appraisal and legitimation of higher-level principles of distribution assessment. The theory of social choice must operate, on their view, not at the local, case-by-case level of specific distributions, but at the global, synoptic level of the general principles of the matter. And they correspondingly insist that, when one approaches the issue from this perspective, the case for the Pareto Principle is by no means open and shut. There is much to be said, for example, for an approach that is to some degree egalitarian. And once one moves in the direction of egalitarian concerns, the Pareto Principle must be abandoned—at any rate in its usual unqualified generality.

To be sure, a supporter of the orthodox economists' approach might perhaps object as follows: "You are counting the Pareto

Principle out much too soon. After all, it too must be construed as a principle of distribution assessment, one that envisages the attractiveness of gains somewhere without losses anywhere. And so why, even on the present "intermediary" approach, should such a principle come to be recognized as playing a predominant, and indeed determinative role?"

The answer to be given here is clear. It is that while this could happen (in some extended speculative sense), there is no good reason to think that it does. For, as we have seen, to put the Pareto Principle into a controlling position we need to maintain that conflicting principles (of distributive justice, egalitarianism, or the like) should be discounted or subordinated to it. And there neither is, nor can be, any good or sufficient reason for so radical a step.

An Economist's Objection

It is easy to visualize an irate economist objecting at this stage along the following lines: "Your discussion overlooks the key reason for a concern with Pareto optimality. Once we abandon the safe ground of a situation where everyone fares better, we come up against the difficulties of having to assess the pros and cons of X's faring somewhat better at the expense of Y's faring rather less well. In short, we come up against the notoriously difficult issue of inter-personal comparisons of utility. Since this problem is intractable we are driven, *faute de mieux,* to some alternative approach that can dispense with any mechanism for interpersonal utility comparisons, and its ability to help us to do so in a decisive advantage of the Paretian approach." However cogent—nay decisive—this objection may sound to a committed partisan of the economist's approach, the issue looks very different from where the moral philosopher sits.

The old-school moral philosopher is not worried about the fact that the members of a society may take a somewhat jaundiced view of one another's utilities—that X, say, may simply not be prepared to grant that Y's "perfectly awful" is really all that bad. The moral philosopher views the prospect of appraising and comparing the utilities of individuals from an external, detached perspective as an ineliminable part of the problem. For him, the overridingly important perspective is that of the external observer (the "impartial arbiter").

And such an arbiter is bound to equate X's "perfectly awful" with Y's, and U's "absolutely splendid" with V's. The impartial arbiter does not ask people to agree on some schedule of interpersonal appraisals, he himself imposes a uniform basis of assessment.[46]

The moral philosopher must be prepared to evaluate utilities; he cannot take them at face value. For him, the crucial point is that what is utility-augmenting may nevertheless prove to be ethically invalid. What counts from the viewpoint of ethical legitimation is not just that people are happier (in a better condition in point of welfare or utility) but how they get that way: by morally healthy and ethically legitimate means (such as sympathy) or by morally reprehensible and ethically legitimate means (such as a perverse schadenfreude). For example, envy may well translate an objective improvement in everyone's condition into a general detriment. In a highly envious society, a step that best advances the happiness of all need not be morally right. But this sort of utility enhancement surely does not deserve moral recognition.

Of course, drawing this distinction among a man's utility interests between those which merit merely de facto recognition and those which merit recognition de jure, between those which deserve an "ideal observer's" recognition and those which do not, has drastic implications for the viability of a utility-based approach to policy assessment. For the orthodox utilitarian must take his utilities as he finds them; he is not free to admit some and dismiss others on the basis of considerations that are, in the final analysis, moral. On a utilitarian approach, normative considerations must reflect considerations of utility and cannot modify or transform them. For the moral philosopher, however, the situation has a very different look.[47]

And so, the economist's insistence that resort to the Pareto Principle has the advantage of avoiding interpersonal utility comparisons leaves the moral philosopher cold. For him, this upshot is not a genuine advantage at all, but rather the claiming of credit for the refusal to do a job that needs to be done.

A Pluralism of Principles

As was suggested above, the position of the old-school moral philosophers envisages a pluralism of principles and criteria of assessment with respect to distributive justice and social rationality. They consider such principles as that of production (amount of good), equality, desert or merit, need, and so forth. The problem, as they see it, is to elucidate the basis and standing of such criteria and to examine the meta-principles that mesh them in systematic harmony in those applied cases where they pull in opposite directions. For since a variety of diverse and so potentially divergent desiderata are at issue, the prospect of conflict and dissonance clearly arises, and there must be a rational mechanism for meshing discordant criteria in concrete applications. The moral philosopher sees the problem in some such terms.

The problem of establishing a suitable meshing within such a plurality of potentially disagreeing principles doubtless is a hornets' nest of difficulty. Some of these are implicit in the preceding critique of the Pareto Principle, which poses the question, Just how should the egalitarian interest be weighed against that of an effectively cost-free advantage to some sector of society? The problems that arise here are complex. The wish to get away from them is only natural under the circumstances. It is certainly understandable why economists, decision theorists, and their congeners are reluctant to get involved and would prefer to settle the matter on the basis of straightforward and "mechanical" considerations such as those of the Pareto Principle. But the fact remains that if the problems being addressed are to be treated properly and to ultimately good effect, and the underlying issues are not to be simply prejudged by undefended presuppositions, then these difficult issues cannot simply be brushed aside.

A Point of Disagreement

Insofar as it is a vehicle for driving home this lesson, a critique of the Pareto Principle affords a vivid illustration of the difference in approach between the social scientist (economist, decision theorist, etc.) on the one hand and the moral philosopher on the other. Now in this connection there is one defensive maneuver to which economists (decision theorists, social-choice theoreticians, etc.) commonly incline that is particularly obnoxious to the moral philosopher. This is their use of the term *social rationality* with the very narrow construction of a prudential pursuit of selfish advantage by the individuals of a social group. There is surely no earthy reason why a perfectly rational person could not devise a perfectly viable rationale for proceeding differently, for example, by rejecting Pareto's Principle as an automatic rule of rational choice in the interests of potentially discordant considerations.

Once again, one comes up against the shortcomings of the concept of economic man and the economists' traditional conception of rationality in terms of the efficient pursuit of prudential self-interest. What the moral philosopher finds particularly objectionable in the proceedings of his colleagues in economics and decision theory is the way they appropriate to their own uses the honorific rubric *rationality*. They enumerate certain principles of assessment— roughly those of atomistically self-interested prudence—and canonize these as axioms of rational decision-making. Such theses are put before us as principles of rationality by fiat or definition or some comparably high-handed act of preemption. We are told with little ado or argument that conformity to a narrowly self-interested modus operandi in choice situations is what necessarily characterizes the choices of the rational man. And, in place of justificatory argumentation, one finds these principles cast as effectively self-evident axioms whose status is somehow foundational, nay virtually definitional, as though such contentions belonged to the very meaning of rationality.

Here, as elsewhere when economists and decision theorists and social choice theoreticians speak so casually of what the rational man does, they manage to conceal under the sheep's clothing of a seemingly descriptive rubric the wolf of a deeply normative commitment, one that is highly dubious and debatable. They arrogate the proud title of rationality from the convenient predilections of their own inherently debatable standpoint. Accordingly, the invocation of rationality by these analysts marks an attempt to secure a dubious and controversial conclusion by verbal legerdemain and to obtain by theft what can only be secured by substantial—and by no means simple—philosophical toil.

Why Save Endangered Species?

7

Man's intellectual, technological, and social progress has greatly enlarged his numbers and improved the quality of his life. But it has done little if anything to enlarge or enhance the scope of nonhuman life on this planet. Quite the reverse! The sabre-toothed tiger is long gone from the face of the earth. The dodo bird became extinct in the eighteenth century. "Lonesome George," the last of the giant turtles of the Abingdon Islands in the Galapagos, is living out a solitary life in our own day. In each case, man's technology has contributed to the unhappy result.

The departure of a species is a particularly grave eventuation in world history. It represents an irreparable loss — an irreversible change in nature that cannot be undone since we must presume that "the genetic material of extinct species cannot be reconstituted."[48] And so, the continued existence of a species has always been viewed by thoughtful men as an especially significant issue.[49] In general, the extinction of a biological species is clearly a matter of justified regret.[50] But there remains the question of why this should be so. Just what rational basis is there for attributing value to natural kinds of things? And there is also the related but distinct issue of exactly why men should strive to avert the extinction of our remote evolutionary cousins whose survival has become endangered.[51]

Species as Bearers of Instrumental and Intrinsic Value

It is tempting and plausible to take a homocentrically self-interested approach, one which sees other species as instrumental goods capable of serving human interests. We humans, it could be argued, have a definite stake in the survival of biological species, and it is this that underwrites the imperative to save them. Such prudential considerations fall primarily into two categories: (1) our *intellectual* interest in species as objects of pleasurable contemplation and as materials for study from which we can learn about the ways of nature;[52] and (2) our *practical* interest in species as vehicles from which (at least potentially) man can extract some material use or benefit, perhaps by using them for companionship or diversion, perhaps by exploiting them for our own economic ends, perhaps simply through their role as useful components of the ecosystem. Such an approach takes the view that we have a prima facie interest in the survival of the sundry species of biological creatures, casting them in the role of instrumental goods — man-relative assets for promoting our intellectual or practical benefit.

As far as they go, such considerations are valid. They do correctly indicate that we have a real and genuine stake in the survival of species. And various important considerations can be accounted for within this framework. For example, it seems not quite so serious a matter if a species dies out while the various others in its close biological neighborhood continue to exist. Or again, it would be seen as not so serious if a species becomes extinct in those rather exotic circumstances where man could "bring it back from the dead" — either synthetically, or by genetic engineering, or by some sort of deep-freeze storage of genetic materials that made its revival at will a routine possibility — so that the extinction is not permanent but reversible at man's pleasure. The fact that we would, under such circumstances, be naturally inclined to regard the extinction at issue as something less than catastrophic is perhaps most readily explained in terms of this homocentrically self-interested view of the matter.

All this is fine as far as it goes. But we can and should go further. For while the homocentrically self-interested posture affords a workable and appropriate perspective, it nevertheless does not represent the end of the matter. The thing cuts deeper. The idea of value is the key here. When a species vanishes from nature, the world is thereby diminished. Species do not just have an instrumental "value for" man; they also have a value in their own right — an intrinsic value.[53]

To be sure, even superficial familiarity with the ways of the world suffices to show that it is man and not the lion who is king of the beasts. Yet whether we look to Genesis or to Darwin, it is clear that man is an afterthought in cosmic history—a lately-come creature who may perhaps crown the effort of creation, but whom we can only with human hubris regard as its raison d'être. It approaches megalomania to view man as the evaluative pivot-point of nature, to see our particular species not only as the *measurer* of value but as its *measure* as well. We can hardly be so parochial as to believe that other biological species exist only for us, so that whatever value they have is man-correlative.[54] It would be more plausible to deny value to nature altogether than to ascribe it solely from the angle of human desiderata and deny that other species can have value in their own right.[55]

At this point, metaphysics becomes germane, specifically that quite distinct mode of metaphysical inquiry which may be characterized as evaluative (or normative) metaphysics. The paternity of evaluative metaphysics may unhesitatingly be laid at Plato's door, but as a conscious and deliberate philosophical method it can be ascribed to Aristotle. In the *Physics* and the *De Anima* we find him not merely classifying the kinds of things there are in the world, but ranking and grading them in terms of relative evaluations. His preoccupation in the *Metaphysics* with the ranking schematism of prior/posterior[56] is indicative of his far-reaching concern with the evaluative dimension of metaphysical inquiry. (A sound insight led anti-Aristotelian writers of the Renaissance, and later preeminently Descartes and Spinoza, to attack the deeply Platonic/Aristotelian conception of the embodiment of value in nature.)[57] A paradigm of this traditional, Aristotelian approach can be found in St. Thomas: "Although an angel, considered absolutely, is better than a stone, nevertheless two natures are better than one only, and therefore a Universe containing angels and other things is better than one containing angels only."[58] The locus classicus of philosophical recourse to metaphysical values is of course its well-known use in the creation-ethic of the system of Leibniz.[59]

The aim of evaluative metaphysics is not to sort (classify, categorize) but to grade (appraise, rank), and its work is not to classify "the furniture of the universe," but rather to assess it. The very possibility of such an enterprise rests on the case for the existence of distinctly "metaphysical" values. They are values in that they are all instances of the application of such evaluative categories as better-or-worse, or more or less significant; they are metaphysical in that they relate to the intrinsic merit of existing things, their "desert for existence." Such values are neither aesthetic (having to do with "enjoyment" in contemplation, primarily in respect to artifacts), nor ethical (bearing directly upon the evaluation of human acts), nor pragmatic (relating to use or consumption or "enjoyment"). They relate to the very being of things in themselves, not necessarily to the realm of human purpose or interests.[60]

According to the old supernatural cosmology, God is the fountain of value even as the monarch is the fountain of honor in a secular kingdom. Once this theocentric viewpoint is abandoned, it would appear that value is left without any independent footing and becomes a human contrivance, a thing of merely human invention or convention. But the choice between God and man, between divine decree and human projection, is a bit facile. There is no reason why one cannot take the stance of a Leibnizian value metaphysic and see the basis for value to reside in the ontological nature of things, to be discovered rather than invented by man. Metaphysical values are neither moral nor theological but sui generis, that is, they simply have their own being and status.

To be sure, nobody would say that rational evaluation in the ontological sphere is easy or even comfortable. "Judge not," the biblical dictum enjoins us, "lest ye be judged." It is simple enough to say that all life is sacred—or at least that there is no inherently worthless species. But this does not remove the difficult issue of "higher" and "lower" forms of life and of their comparative evaluation. This will doubtless turn largely on the magnitude of the capacities and potentialities of the organisms at issue: "If we prefer to save the life of a fellow-man rather than that of the organism responsible for yellow-fever—an organism no doubt of considerable interest and beauty—it is because he has potentialities the yellow-fever organism lacks, potentialities for evil, admittedly, but also for good. . . . The yellow-fever organism cannot love, for example, or exhibit courage, or create works of art. It does not suffer as human beings suffer, or live in fear of death."[61] It is worth heeding Bertrand Russell's useful reminder that when we deem ourselves superior to the amoeba, it is *we*, and not the amoeba, who espouse this view. But then, too, view-espousal is something which the amoeba cannot do at all. One does well to agree with Stuart Hampshire here: "The peculiar value attached to human life . . . [is] not dependent on regarding and treating human beings as radically different from other species in some respects that cannot be specified in plain, empirical statements . . . [but] the exceptional value attached both to individual lives, and to the survival of the [human] species as a whole, resides in the power of the human mind to begin to understand, and to enjoy, the natural order as a whole."[62] The value of human life derives from the value of what it can bring to realization in nature. Why should the case of other species be all that different?

Let us consider how the concept of obligation bears on these metaphysical values. The fact that species have an intrinsic value in their own right, and not merely an instrumental value-for-man, means that their conservation is not just a matter of prudence but one of duty as well. But what sort of duty?

It is not a moral duty.[63] Moral obligation is correlative with two injunctions. First we have the negative imperative, *neminem laedere*, "Do not injure the legitimate interests of others." Secondly there is the positive imperative, "Promote the legitimate interests of others — enhance the realization of *their* good." Moral obligation is thus always interest-oriented. But only individuals can be said to have interests: one only has moral obligations to particular individuals or particular groups thereof.[64] Accordingly, the duty to save a species is not a matter of moral duty toward it, because moral duties are only oriented to individuals. A species as such is the wrong sort of target for a moral obligation.[65]

Nor is a moral duty to the *members* of other species at issue. For moral duty toward individuals involves a reciprocity of mutual communal involvement that — as matters now stand — does not cut across the boundaries of the "family of man," thanks to the absence of sufficiently tight bonds of community on other fronts. (This is a difficult issue which must rest here simply on the basis of a dogmatic fiat.)[66]

To be sure, the moral aspect is not entirely irrelevant to the saving of endangered species. But insofar as it is a moral duty that is at issue, this moral duty is one we owe to our fellow men. Men have various legitimate prudential interests in the preservation of species (as we saw above). And this yields a straightforward route to a moral obligation. The duty at issue resided in the requirement to safeguard the right (the "birthright") of men—even of those yet unborn—to have a rich, diversified, and interesting environment. The moral duty at issue is thus our obligation to other men for preserving species in view of the very real stake that we have in their existence.

This line of thought at once raises the question of whether this human stake in the continued existence of biological species is the end of the matter. Is the only duty we have to save them a duty to our own fellows? Surely not. A very different sort of duty is also operative, an ethical duty in a broader sense than the specifically moral one—whose basis is larger than the specific interest-orientation of more narrowly moral obligation. It is "higher," "nobler," and more "disinterested" than a specifically moral obligation. Moral duty is aimed at safeguarding the interests of others. Ethical duty is oriented more generally at the enhancement of value in the overall existence of things.[67]

Ethical obligation has two forms, parallel to the case of moral duty. There is first a negative duty to preserve: do not act to the detriment of what has value—do not diminish the sphere of value. Second, there is a positive duty to enhance: act so as to protect and promote what has value—augment the sphere of value. The fundamental obligation to promote the good—to act so as to enhance the realization of values within nature—is an ethical obligation in a sense that goes beyond the domain of specifically moral obligations. Ethics in this sense is not limited by the traditional horizons of morality, confined to the preservation of human inhabitants and the enhancement of their lives. Not only can it concern itself, as "evolutionary ethics" has done, with the preservation of "the" (i.e., our) species, but with that of others as well.

With respect to this ethical mode of obligation to save species, we may deploy a syllogism based on two premises: (1) species are (ceteris paribus) bearers of value—not just instrumentally, but metaphysically; (2) we have a prima facie obligation—an ethical obligation in the broadest sense—to promote the enhancement of value. And this leads to the conclusion that it is a matter of duty (of ethical duty in the broad sense) to further the realization of objects of value: in this case, species.

This brings the discussion to the end of one line of thought with regard to our basic question: Why save species? The answer developed here runs as follows. We must do this for three reasons: (1) on prudential grounds, because we ourselves have an interest in them; (2) on moral grounds, to safeguard the legitimate interests of other men; and also (3) on ethical grounds, in view of their embodiment of metaphysical value in their own right.

Our Duty to Save Species Does Not Derive from Their "Right to Exist"

It is sometimes said that biological species have a claim on our protection because they have an inherent right to existence. We are told: "If you allow another species to become extinct, then you infringe upon its rights." But this conception rests on a category mistake. For only individuals can have rights or claims, not species, classes, categories, or such. The mistake arises from the common misconception that duties are correlative with rights, as, for example, that my duty not to steal your property derives from your right to its possession, or that my duty to perform my work satisfactorily arises from your right as my employer to demand this of me. But this approach is overly legalistic and contractualistic. I have a duty to endeavor to save the drowning man, but he does not have a right to my assistance. I have a duty to develop my human potential where I can, but this does not mean that anyone has a right to expect this of me. We need not go about inventing rights to validate duties and need not deny the existence of duties in the absence of rights. Thus animals themselves are not moral agents and so have neither duties nor rights; but that does most emphatically not mean that men do not have duties toward them—*humanité oblige* is a perfectly tenable precept.[68]

John Henry Newman wrote: "We have no duty towards the brute creation; there is no relation of justice between them and us. . . .

[They] can claim nothing at our hands."[69] This gets the matter only half right. To be sure, animals have no claims or rights vis-à-vis man—no "relation of justice" obtains.[70] But this fact does not preclude men from having duties toward them.[71] We are not responsible to nature (or its constituent species and individuals), but this does not mean that we are not responsible for it (or for them). Thus as regards their claims, Newman was quite right—at any rate as far as biological species are concerned: a species does not have claims on us for its consideration. Yet this is not the whole story. For there certainly is a claim—an impersonal claim implicit in the nature of things—for our consideration of a species. It is necessary—ethically, morally, and prudentially—for us to be considerate of species. The locution "There is a claim for our consideration of species S" is valid. But the very different claim "Species S has a claim on us for consideration" is literally false, and it is figuratively true only insofar as it can be glossed as a (misleading) formulation of the former thesis.

The ethical duty to preserve species thus emerges as a quintessentially humanitarian task. It roots in our ethical duty to promote value. And it looks to no reciprocity or return in terms of any prudential human interests—apart from our essentially spiritual, or ethical, interest in the existence of value. (But this broadly ethical interest is not a matter of theology—it does not relate to our obligations to God or to anyone apart from "what we owe it to ourselves to do" by way of the cultivation of human excellences.)[72]

It must, on the other hand, be recognized that this question of rights does bear on the issue—but in quite a different way. The key fact here is that from an ethical point of view we have no right to allow another species to come to grief needlessly. For—quite in general—there is inherent in the very logic of the conceptions of rights and duties a fundamental principle to the following effect: If X has no right to do A, then X has an obligation (duty) not to do A.[73] And so the fact that we have no right to allow another species to become endangered entails that we have a duty to do what we can to avert threats to its existence.[74] The crucial consideration here lies not in the possession by others of certain rights, but in the lack thereof on our own part.

The Duty to Save Species Is Defeasible

It must be recognized and stressed that the ethical obligation to protect species is a prima facie obligation — one that is not absolute and all-overriding but defeasible in the light of countervailing considerations of sufficient weight. Both from the prudential and the ethical standpoint, it is not an absolute good that a species should continue to be exemplified, but only a good when other things are equal (ceteris paribus), that is, only when these other issues are suitably favorable. For example, in circumstances where a species that stands "higher" (in the order of appraisal of metaphysical valuation — presumably in terms of its overall repertoire of capabilities) is imperiled by the continued existence of a lower species, it will be permissible (and presumably in some circumstances even a matter of duty) to endeavor to eradicate the latter.

One somtimes finds intrinsic mistaken for absolute, rival-overriding, value. For instance, John Passmore has written: "We can think of wilderness and of species as having either a purely instrumental or an intrinsic value. On the first view, wilderness and species ought to be preserved only if, and in so far as, they are useful to man. On the second view they ought to be preserved even if their continued existence were demonstrably harmful to human interests."[75] But this does not get the matter quite straight. Intrinsic values too can be outweighed by other intrinsic values — they can be prima facie rather than absolute. The intrinsic value of a species can be outweighed by the intrinsic value of an enhanced condition of human well-being. Here as elsewhere it is a matter of weighing values against one another.

Our prudential interest in the continued existence of one species can be overridden by a greater prudential interest in the continued existence of yet another that stands in conflict with it. This point is obvious and needs no development. And the ethical duty to save a species is similarly defeasible. Our prima facie duty to safeguard other species does not mean that we could not justifiably eradicate a noxious species of microorganisms, for example, or a highly deadly species of insect.[76] In particular, the right of self-preservation is fundamental and can — in suitable circumstances — override other duties that conflict with its exercise. It would in certain cases be a wholly justifiable endeavor to eradicate all members of a man-endangering species and to assert our right against theirs. Nor is self-preservation the only possible counterweight to our duty to act for the preservation of species. The interest we have in other species can be overridden by other interests; the duty to preserve them can be outweighed by other duties. The realization of some larger advantage of the welfare of our fellows — the moral duty we have toward other men for insuring their welfare — may well make the extinction of a species permissible. Nor need it be our species that is at issue. To assure one good, it may be necessary to consign another to destruction.

It is, of course, possible to argue that it is always our flat-out duty to endeavor to save any and every species and that no considerations, interested or disinterested, can countervail against this altogether indefeasible duty. But this "whatever is, is right" posture has little to recommend it.[77] It seems implausible to hold that the extant arrangement is the best possible and that prudential (man-oriented) or ethical considerations are in all circumstances insufficient ever to indicate the justification — perhaps even the duty! — for extinguishing a biological species. To take this stance would be to endow the vagaries of the evolutionary process with a weight that scarcely seems plausible given our understanding of its modus operandi.

Still, the loss of a species is never something to be taken lightly. It is a matter of complex and painstaking cost-benefit analysis with all due safeguards because of the decisiveness and finality of the transaction. In his poem, "Song of the Redwood Tree," Walt Whitman asserts that the redwood must "abdicate" his forest-kingship so that man can "build a grander future." This is a very dubious proposition. For there is good ground for holding that a human future without the redwood is eo ipso less grand, that man's interest in its preservation is substantial, and that our duty to insure its continued existence for the benefit of our fellows outweighs the counterconsiderations.

A Question of Priorities

In addition to the problem of weighing species against species, we encounter the difficulty of allocating scarce resources, which pits the saving of species against other, specifically human interests and responsibilities. For there is no question that the saving of species can be expensive. In some instances, preservation in zoos may be the best prospect, and zoos are costly institutions. In other cases human intervention "in the field" on a considerable scale may be needed to preserve a species in its natural habitat. Protecting certain species of animals by avoiding the use of pesticides and herbicides may exact a substantial price in terms of a lower standard of diet for local human populations.

And so we face that crass economic question: How much is it worth? And in the present context this question has deep philosophical ramifications. To what extent, for example, is the multiplication of the number of persons living, and the enhancement of the quality of their lives to take precedence over the maintenance and perhaps even the elevation of the level of nonhuman existence? What Charles Hartshorne has written on this issue makes good sense:

To risk a man's or woman's life for a subhuman individual is, I believe, unwarranted. But to do so to save an entire species, say of whale, ape, or elephant, would this be unwarranted? I'm not so sure. In the biblical tradition the other animals were said to be there for the sake of man. Still, cruelty to animals was frowned upon, and in Job and in the sayings of Jesus there seems to be a feeling that the nonhuman creatures have their own values, apart from any utility they have for us. "Ye are of more value than many sparrows," but this implies that they too have some value. No form of life should be thought a mere means, a mere utility. All forms are beautiful and good in themselves. Here I agree with Schweitzer against Kant.[78]

The resources we expend in saving a species are resources we cannot commit to saving or improving human lives—to hospitals and well-balanced school lunches, to medical research and the seemingly unwinnable "war on poverty." A very difficult cost-benefit comparison arises in such cases. We can do something—but only, alas, very little—with the specifically homocentric bearing of our interest in the preservation of species. For, as we have seen, the interests we have in species are in large measure intellectual rather than practical and welfare-oriented; they relate largely to knowledge, beauty, life enrichment, and so forth, rather than to actual utility. The human interests involved are real but generally not very urgent. Our actual stake in species will often relate primarily to excellences which are in fact of actual interest only to a few. And the weight of the duties engendered by considerations of metaphysical value are even graver imponderables.

One important consideration must, however, be borne in mind here. The extinction of a species is in general an irreversible development. Now in reckoning the price we pay for such an eventuation, we must bear in mind that there will not only be the recognized costs that can be calculated in advance, but there may also be unforeseeable costs which — given the incompleteness and inadequacy of our knowledge of how things work in nature — remain altogether unforeseen. In the face of irreversible changes the presence of such imponderable risks should always give us pause. The historical evidence does not indicate that we are in a position to advance impressive claims to understand the "balance of nature."

The Issue of Creating New Species

The whole of the preceding discussion concerns the issue of protecting and preserving existing species. The question remains whether our ethical duty to save existing species carries over to a duty to create new species when we have the capacity to do so through eugenics or genetic engineering or whatever. Recent advances in nuclear biology, for example, have made the prospect of creating new microorganisms a realistic one with very far-reaching implications.

This delicate and problematic issue of our duty to bring new species to realization also has both a prudential and an ethical dimension. As regards the former, it is clear that we can have exactly the same sort of prudential interest in the creation of a new species as in the preservation of an existing one. Moreover, the ethical injunction "Enhance value!" cuts the same way in both cases. To all appearance, the normative considerations run parallel on both sides, and the issue of saving the old and creating the new stand on exactly the same evaluative footing. Accordingly, it would seem that the duty to save existing species carries over to an analogous duty to create new ones.

Yet this view of the matter, though at bottom correct, is somewhat simplistic. The key difference between saving an extant species and creating a "merely possible" one is more practical than theoretical. An extant species is a known quantity, and to the very extent that it is itself endangered it is presumably no threat to others. But creating a new species is always a leap in the dark that brings a new and largely imponderable situation to realization. In Hamlet's words it exchanges "the evils that we know for those we know not of." The difference at issue is in the final analysis a matter of prudence—of averting what could (for all that our imperfect intelligence can tell) prove to be a real threat to our own position in the scheme of things.

To be sure, both preserving old species and creating new ones are are simply ways of intervening in the evolutionary course of events to produce conditions that would otherwise not be. But there is a drastic disparity between preservationism and creativism. Our obligation to protect the extant is of a different order from that of optimizing the merely possible. The fact that some species is in actual possession of an ecological niche—it is already there—carries substantial ethical weight in its own right. The value of some purely possibilistic alternative cannot countervail against it. (This, surely, is one of the lessons of evaluative metaphysics.) For one thing the "balance of nature" has already had the chance to come to terms with an existing species, whereas a new species creates a new and imponderable situation. And the balance of nature is in general extremely delicate; any disturbance of it, however slight, is likely to have massive repercussions. Given the "purely practical" consideration of this delicate balance—the fact that in ways perhaps difficult to foresee in detail but almost inevitably real, the generation of new species would or could upset the balance and endanger our own existence and that of other extant species—that our presumptive duty to create species on the basis of the ethical obligation to maximize value will by and large have to remain in suspension as subordinate to the claims of self-preservation. By contrast, the very fact that an existing species has become endangered indicates that its threat to the balance of nature is altogether minimal. No doubt, if the creation of new species could be achieved in conditions of absolute safety— say by operating through remote control in a distant and otherwise unpopulated galaxy—then the force of the injunction "Enhance value!" would become operative in this creationist sphere as well. But apart from this rather far-fetched situation, there is a significant asymmetry between conservationism and creationism with respect to biological species. However, the asymmetry at issue here emerges in the final analysis as a practical matter of prudence rather than one of strictly ethical or metaphysical considerations.

Scientific Progress and the "Limits of Growth"

8

One often hears critics lamenting that the advent of "big science" has brought such evils as inefficiencies of scale, administrationitis, a dampening of initiative, and the public relations promotions needed to obtain funding.[79] For example, Sir Karl Popper in a recent publication surveys with great dismay the economic involvements of modern science. He casts them in the role of hindrances to scientific progress and laments that "Big Science may destroy great science."[80] The labors of big science, we are told, simply get in the way: "Further progress will come . . . not from more and more measurements with increasingly powerful machines, but from a revolution of thought. . . . In place of the bizarre but apparently meaningless phenomena reported every few months [in high-energy physics] . . . we need a new Einstein to bring order by new intellectual analysis."[81] What is needed, according to many, is less big science and more high thinking. This perspective strikes me as badly mistaken. For there is good reason to view the onset of big science as not a regrettable aberration but an inevitable fact of life throughout the natural sciences.

First Things First: Technological Escalation

Nature inexorably exacts a dramatically increasing effort of enhanced sophistication in data deployment for revealing her "secrets" and accordingly becomes less and less yielding to the efforts of our inquiry at given fixed levels of information-gathering technique.[82] In natural science we do the easy things first. The very structure of scientific inquiry, like an arms race, forces us into constant technological escalation where the frontier equipment of today's research becomes the museum piece of tomorrow under the relentless grip of technical obsolescence. It would thus be as futile to follow various writers into hankering after the days of "string and sealing-wax" apparatus as it would be to join Talleyrand in lamenting the lost *douceur de la vie* of the old regime. There is no point in blaming human foibles or administrative arrangements for a circumstance that is built into the very structure of investigation in natural science (realizing, to be sure, that the facts that make bigness a necessary condition of significant progress do not establish it as sufficient). The enormous power or sensitivity or complexity deployed in present-day experimental science has not been sought for its own sake, but rather because the research frontier has moved into an area where this sophistication is the indispensable requisite of ongoing progress.

In any problem area of a matured branch of natural science, continually greater capabilities in point of "capacity" are required to realize further first-rate results. Once all the findings accessible at a given state-of-the-art level of investigative technology have been realized, one must move to a more expensive level. An ongoing enhancement in the quality and quantity of input becomes requisite — one requires more accurate measurements, more extreme temperatures, higher voltages, more intricate combinations, and so on. As the range of telescopes, the energy of particle accelerators, the power of neutron-beam generators, the effectiveness of low-temperature instrumentation, the potency of pressurization equipment, the power of vacuum-creating contrivances, and the accuracy of measurement apparatus increases — that is, as our capacity to move about in the parametric space of the physical world (as it were) is enhanced — new phenomena come into our ken, with the result of enlarging the empirical basis of our knowledge of natural processes. The key to the great progress of contemporary physics lies in the enormous strides that have been made in this regard; as one commentator has justly observed: "In almost every observational dimension, short time, small distance, weak signal, and the like, the limits are being pushed beyond what might have been reasonably anticipated."[83]

The idea of "new phenomena" is the talisman of contemporary experimental science — and of physics above all. Workers in this domain do not talk of increased precision or accuracy of measurement in their own right; it is all a matter of the search for new phenomena. And the development of new data technology is virtually mandatory for the discerning of new phenomena. The reason is simple enough. Inquiry proceeds today on such a broad front that the range of phenomena to which the old data technology gives access is soon exhausted. And the same is generally true with respect to the range of theory testing and confirmation that the old phenomena can underwrite.[84]

A look at the history of major scientific innovation indicates that it is generally not a matter of spontaneous generation but rather a provoked response to three sorts of challenges that crop up uninvited and often unwelcome. We regularly encounter situations in which the equilibrium of "established" theory and "familiar" fact is upset by the accession of new facts, creating a situation in which new data sources are at issue because existing theories can presumably—and usually—accommodate the old data (this, after all, being the basis on which they are the "existing" theories).[85] The new data or data complexes create a disturbance of the cognitive equilibrium in the general coherence of the overall constellation of data-cum-theory. A setting is thus created—a novel problem-situation—in which scientific innovation (i.e., readjustment on the side of the accepted scientific theories) becomes a situational imperative.

It is not so much that new phenomena indicate the need for innovative theorizing, though this is indeed generally the case. But also, for a new theory to establish itself as duly progressive—as representing an advance on which in turn yet further steps can and should be based—this theory must afford some guidance in explaining hitherto unknown phenomena, phenomena which can only be attained by a more powerful technology. An equilibrium between the old phenomena and the old theories was presumably established long ago; the anomalies that demand theoretical innovation generally come from the new phenomena afforded by novel technology.[86] Thus as regards this crucial issue of *substantiation*, the matter can be viewed from the familiar perspective of the well-known model of science as an enterprise of framing and testing explanatory theories (in Popperian terms, the model of conjecture and refutation). It is clear that one will here soon exhaust the tests available within the familiar sector of the parametric range and must move on to virgin territory to carry out the further tests on which (on his model) theoretical progress is crucially dependent. The cleverness of theoreticians assures that the data attainable over the old physical parameter-range will soon come to be accommodated, and their refutatory prospects exhausted, so that one must press on into heretofore inaccessible regions of power or complexity. For at each stage of scientific inquiry we face the need to reduce a diversified spectrum of live hypotheses to manageable numbers—a need that grows more elaborate with the exponential proliferation of possibilities that lies down the road of sequential conjecture, with iffy hypothesis piled upon iffy hypothesis.

The Exploratory Penetration Model: An Arms Race Against Nature

Scientific progress thus depends crucially and unavoidably on our technical capability to push outward in ever widening circles into the increasingly distant—and increasingly difficult—reaches of the power/complexity spectrum of physical parameters to explore and explain the ever more remote phenomena to be encountered there. It is of the very essence of the enterprise that natural science is forced ever further into the extremes of nature.

The assumptions that underlie the detailed functioning of such a prospecting model of inquiry in natural science may be set out in terms of three theses.

1. The processes of nature hinge on the interconnected operation of certain parameters, each such parameter varying over a range which defines one particular physical dimension of the "phase space" whose exploration is the task of natural science. (These parametric ranges are such that in the case of an inherently limited range—e.g., speeds to the velocity of light, temperatures to absolute zero—an "order of magnitude" step along the range carries us yet another one-tenth of the way to the limit, as per the sequence: to within 10 percent of the limit from "where we started from," to within 1 percent of it, to within .1 percent of it, etc.)

2. We humans ourselves occupy a certain (for us) natural position within this parametric space—a location within it that is the (for us)

natural starting point for its exploration. (For example, our sensory organs respond to radiation in the visible part of the electromagnetic spectrum or in the audible range of vibration frequencies.) In our own "parametric neighborhood" we can, thanks to the equipment of our evolutionary heritage of perception, generally make explorations relatively simply and cheaply.

3. Moreover, by suitable technical means we can explore regions of this phase space at locations increasingly remote from our natural starting point. This is an ongoing process and one which is altogether crucial to scientific progress because the cognitive potential of fixed-distance explorations is soon exhausted. We are forced to move ever further outward in parametric remoteness or complexity of detail. This requires a rapidly increasing commitment of effort and resources. It is as though we needed to overcome the impedance of a restraining force which pulls us toward home base and—unlike gravitation—makes it increasingly difficult for us to move further away from it. To pursue the metaphor of a voyage of exploration, the further the journey from home base, the larger the vessel we need and the more difficult and complicated is its outfitting.

This view of the matter involves no denial of the reality of the process of technological leapfrogging by breakthroughs to increasingly sophisticated levels of technological progress. It is just that this growth in effectiveness produces no savings, but—quite the contrary—is ever more expensive. In natural science we are involved in a technological arms race: with every "victory over nature" the difficulty of achieving the breakthroughs which lie ahead is increased. One is simply never called on to keep doing what was done before. A Sisyphyus-like task is posed by the constantly escalating demands of science for the enhanced data to be obtained at new levels of technological sophistication. One is always forced further up the ladder, ascending to ever higher levels of technological performance—and of costs.

Technological Escalation in Science: The Unavailability of Economics of Reproduction

The preceding considerations set science apart from productive enterprises of a more ordinary sort. The common course of experience in manufacturing industries since the industrial revolution yields a historical picture where (1) the industry has grown exponentially in the overall investment of the relevant resources of capital and labor, whereas (2) the output of the industry has grown at an even faster exponential rate. As a result of this combination, the ratio of investment cost per unit of output has declined exponentially due to the favorable "economies of scale" available throughout the manufacturing industries. However, this relationship does *not* hold for the science industry — the "industry" of the scientific enterprise itself.

The economies of mass production are unavailable in research at the scientific frontiers. Here the existing modus operandi is always of limited utility — its potential is soon wrung dry as the frontiers move ever onward. Of course, if it were a matter of doing an experiment over and over, as in a classroom demonstration, then the unit cost could be brought down and the usual economies of scale would be obtained. The economics of mass *re*production is altogether different from that in pioneering production. With mass production costs, it is as though each item made chipped off a bit of the cost of those yet to be made. Here too an exponential relationship obtains, but one of exponential decay.[87] To draw a somewhat odd biological simile, it is as though the initiation of the species was expensive, but once the species is realized the further generation of its individual members becomes increasingly simple. The situation with the research and development costs of scientific research technology is altogether different.

Frontier research is just that — true pioneering. What counts is not doing it, but doing it *for the first time*.[88] The situation of the initial reconfirmation of claimed discoveries aside, repetition in research is in general redundant and thus usually pointless. As one acute observer has remarked, one can follow the diffusion of scientific technology "from the research desk down to the schoolroom."[89] The savings of mass reproduction in science are thus useless outside the context of instructional purposes. In innovative science there are no economies of scale. And these economic realities of course betoken economic limits and limitations. The finitude of our resources must be realized and reckoned with.

Implications of Zero Growth

The existence of such economic limitations upon inquiry poses a fundamental question about the extendability of our knowledge of the world. Is there in fact any reason to think that the physical infeasibility of data access to certain remote sectors of parametric space will seriously impede the expansion of our scientific understanding? After all, if for some reason certain parts of the earth could not be explored, this merely geographic deficiency in our information would almost certainly not retard in any substantial way our knowledge of the earth's structure, history, or geology.

But this analogy is misleading. With natural science it is not just some minor and venial deficiency that is at issue. The regions in which our access to data is blocked by physicoeconomic limits are precisely the regions where certain peculiar and characteristic sorts of natural processes take place, processes we have every reason to regard as crucial for our understanding of nature. The sector of parametric space we are prevented from viewing is not just another random region, information about which we could pretty well by-pass by extrapolation from without; by its very nature it is a region whose phenomena are—or may, in the face of all past experience, safely be presumed to be—different in kind from those to which we already have access.

At this point it is necessary to introduce an important distinction. Scientific problems can be of two kinds. (1) There are *synthetic* problems which are holistic rather than factorable and have a gestalt unity in virtue of which they do not admit of decomposition but must be resolved all at once through the deployment of a suitably powerful effort. (2) There are *analytic* problems which can be factored into pieces admitting of stepwise resolution, so that while a vast total effort may be required for the solution of the problem, this can be expended seriatim in small installments. (A hickory nut can be cracked only by concurrently marshaling a sizable effort; a ball can be rolled to a far-off destination by tiny bits of effort expended successively.) Problems in the first category may be characterized as *power-intensive*, those in the second as *complexity-intensive*.[90]

The investigation of fundamental particles through collisions brought about by high-energy particle accelerations typifies the category of power-intensive problems. The etiology of cancer and cardiovascular disease affords a prime instance of complexity-intensive research problems, perhaps due to the convoluted functioning of cellular mechanisms. (For example, it is often impossible in the present state of knowledge to predict whether a deliberately induced immune response will stimulate or inhibit the growth of a cancer.) Environmental biology (ecology) with its proliferation of delicately interconnected variables, interlocked like a jigsaw puzzle in intricate configurations, also exemplifies this complexity-intensive case. Or again, molecular biology, with its thousands of investigators painstakingly elucidating step by step the mode of action of the nucleic acids and other cellular constituents typifies the category of complexity-intensive problems.[91]

But the perfectly realistic prospect must be faced that there will be some problems in natural science that are inherently power-intensive — problems for whose resolution one needs to have more or less "simultaneously" a nonaffordable quantity of nonstoreable resource inputs. In such cases, the escalating level of resource demands, such as the size of equipment or the magnitude of energy requirements, will eventually pose a practically insuperable obstacle. Here we ultimately approach, in a zero-growth world, a decisive de facto limit in the attainability of data and hence in the realizability of results, a limit not of possibility but of feasibility. The resolution of these problems becomes insoluble not because they are inherently so, but primarily because those technical measures indispensable to their resolution cannot be put at our disposition within the limits of available resources.

To be sure, the situation is quite different in the analytic, complexity-intensive domain where the analytic aspect of complication/compilation/complexity can always be accommodated sequentially. Unlike synthetic problems, analytic issues are always factorable; they can be decomposed into smaller pieces that can be resolved "on the installment plan." Thus with the analytic, complexity-intensive problems of the power-attainable domain, only the sky (or, speaking less figuratively, the horizon of time and the reach of human patience and ingenuity) is the limit.

The existence of this category of complexity-intensive problems is accordingly critical for the long-run prospects of scientific progress. With the data of the synthetic (power-relative) domain, an effectively insuperable economic barrier must eventually be reached.[92] But with analytical problems such a barrier can always in principle be bypassed, given sufficient patience.[93] Sufficiently elaborate exploration within the "already accessible" parametric regions can always yield more sophisticated insights into the complexity-levels of physical processes.

The limits that operate in a zero-growth world with respect to our ability to dedicate increasingly massive resources to the power-intensive areas of scientific investigation have far-reaching consequences. Inaccessible phenomena entail inaccessible findings. If physical limits and their economic amplifications restrict the scale of our interactions with nature, they thereby limit our cognitive access to the phenomena of nature, and therefore—so we cannot but believe—they restrict access to its laws as well. Data restrictions are bound to compromise the adequacy and completeness of our consequent theorizing. The standard empiricist posture must prevail: our knowledge of the world is empirical, that is, data-dependent, and any significant deficiency in our access to phenomena must be presumed to entail a correlative deficiency in an understanding of the natural laws that govern them. Our knowledge of the world accordingly becomes circumscribed in such a manner that we must accept the humbling lesson of Hamlet's dictum that "There are more things in heaven and earth, Horatio." The existence of knowledge-impeding limitations on access to the phenomena must be acknowledged to have decisively negative implications for the claims of science to the realization of adequacy and comprehensiveness.

The cruel fact is that in a zero-growth world in which the resources dedicated to science at any given time are stable (rather than growing exponentially as they have done in the past) there will inevitably be a substantial retardation in the pace of progress because of the escalating costs of work at the scientific frontiers.

The Technological Imperative

One oft-encountered objection to the idea of limits to scientific progress goes as follows: "All this attention to limits of inquiry and boundaries of knowledge is gloom and doom thinking. Do not sell human intelligence short! Man's intellectual creativity will ultimately prevail to overcome all these apparent obstacles to smooth scientific progress. What looks like an insuperable limit from our present-day perspective may well crumble down under the next breakthrough that lies just around the corner."[94]

But this is surely wishful—and highly unconvincing—thinking. The obstacles that have been canvassed in our discussion are technological limits on data availability which are themselves inherent in the very nature of the physical world, as best science itself depicts it for us. Natural science is empirical science; it indispensably needs data. Only a rationalist of an extreme and doctrinaire stamp could think that sheer intelligence can surmount data barriers in mastering the "secrets of nature." To hold that pure thought and unaided intellect can overcome the physical limitations correlative with data dependency is alien to the very nature of empirical science. (However great its reliance on imagination and creative thought, natural science is something very different from imaginative literature or pure mathematics, where creative force of intellect can operate alone, without the constraining restrictiveness of material limitations.) These limits manifest themselves to us in a fundamentally economic manner. Our physical interventions in the course of nature are everywhere subject to the limitation of economic constraints upon the availability of resources. There is, and can be, no adequate reason to think the case is otherwise in the domain of empirical inquiry, where our interaction with nature is a driving force of the enterprise.

The position presented here is thus not a matter of "gloom and doom thinking" at all, but simply one of being realistic — of facing the facts as best we can make them out. It accepts squarely and honestly the inexorable implications of the economic limits upon the progress of inquiry in a context where the deployment of technology in the acquisition and the processing of data represents an unavoidable fact of life.

On Not Excusing Inadequate Efforts

One important final point deserves stress. Even if it were certain that mankind had pushed its effort at scientific inquiry deep into the region of diminishing returns, one should not suppose that this had gone too far. The evaluation of the costs of scientific knowledge in terms of material and intellectual resources must be offset by recognition of the benefits of scientific work. And these benefits should not be construed in the narrowly utilitarian, gadgetry-oriented sense of *panem et circenses*. It is important to recognize that not just material but intellectual benefits are involved. For while scientific progress has indeed produced an immense benefit in terms of physical well-being, there remains the no less crucial fact that it represents one of the great creative challenges of the human spirit.[95] Man's intellectual struggle with nature deserves to be ranked as a key element of what is truly noble in human life, together with our social efforts at forging a satisfying life-environment and our moral strivings to transcend the limitations of our animal heritage. The recognition of human limits and limitations affords no excuse for inadequate efforts. The runner realizes he will not run a one-minute mile; the thrower recognizes that he will not hurl his discus for one thousand yards. But this does not excuse him for failing to do his best or prevent him from striving to do it. And there is no good reason why the situation in natural science should be any different.

A society that spends many billions of dollars on a varied cornucopia of deleterious trivia, to say nothing of untold billions on military outlays, assumes an uncomfortable moral posture in deciding that science — even big and expensive science — is a game that's not worth the candle.[96] The scientifically and technologically most advanced countries today spend some 2 to 3 percent of their GNP on research and development and some 3 percent of that on basic science.[97] This allocation of roughly one-tenth of one percent of GNP to pure science is certainly not exorbitant and perhaps not even seemly, considering the size of our material and intellectual stake in the enterprise.[98]

Notes

1. J. B. Bury, *The Idea of Progress* (London, 1920), pp. 332–33.

2. *Leibniz: Selections*, ed. P. P. Weiner (New York, 1951), pp. 584–85.

3. R. H. Lotze, *Microcosmus*, trans. E. Hamilton and E. E. C. Tines, 2 vols. (Edinburgh, 1885–86), II, 396.

4. For example, Gallup used "fairly happy" for the middle group while NORC and SRC used "pretty happy."

5. The discussion in this section and its immediate successor draws on the author's book *Welfare: The Social Issues in Philosophical Perspective* (Pittsburgh, 1972), chap. 3.

6. Judgments of this sort, even about oneself, are notoriously problematic: "It is hard enough to know whether one is happy or unhappy now, and still harder to compare the relative happiness of different times of one's life; the utmost that can be said is that we are fairly happy so long as we are not distinctly aware of being miserable" (Samuel Butler, *The Way of All Flesh*).

7. Sentiments of this tendency are readily found across the entire political spectrum.

8. A. Campbell, P. E. Converse, and W. L. Rodgers, *The Quality of American Life* (New York, 1975), make use of this Epicurean formula and offer support for it in chapter 6. Another empirical study regarding this bit of speculative philosophy as to the relationship between expectation and (probable) achievement is Arnold Thomsen, "Expectation in Relation to Achievement and Happiness," *Journal of Abnormal Social Psychology* 38 (1943), 58–73. Other related discussions and further references are given in James G. March and H. Simon, *Organizations* (New York, 1958); Richard M. Cyert and James G. March, *A Behavioral Theory of the Firm* (Englewood Cliffs, N. J., 1963); T. Costello and S. Zalkind, *Psychology in Administration* (Englewood Cliffs, N. J., 1963), esp. pt. 2, "Needs, Motives, and Goals." It is worth noting that often one finds "aspiration" in place of "expectation" in the denominator of the basic proposition. The difference is important but subtle. The enterprising person may aspire to more than he expects to realize; the all-out optimist may expect to realize more than what he aspires to.

9. Rousseau's *Emile* works this line of thought extensively: "True happi-

107

ness consists in decreasing the difference between our desires and our powers in establishing a perfect equilibrium between the power and the will. . . . misery consists not in the lack of things, but in the needs which they inspire." *Emile,* trans. B. Foxley (London, 1911), pp. 44–45. I owe this reference to Peter Hare.

10. An important lesson lurks in this finding, to wit, that consideration of only the idiosyncratic happiness of a society's members is a poor measure of its attainments in the area of social welfare. It would only be a good measure in a society whose expectations held fairly constant or, if not that, at least developed in a "realistic" manner, that is, in a gradualistic pattern that did not automatically leap beyond increasing attainments.

11. For an interesting discussion of cognate issues see Philip Brickman and Donald T. Campbell, "Hedonic Relativism and Planning the Good Society," in *Adaptation Level Theory,* ed. M. H. Appley (New York, 1971), pp. 287–302. One of the interesting points of this discussion is its conclusion that "there may be no way to permanently increase the total of one's pleasure except by getting off the hedonic treadmill entirely" (p. 300).

12. For an interesting discussion of some relevant issues see Philip Brickman and R. J. Bulman, "Pleasure and Pain in Social Comparison," in *Social Comparison Processes,* ed. J. M. Sols and R. C. Miller (Washington, 1977), pp. 149–85.

13. From Joseph F. Coates, "Technological Change and Future Growth: Ideas and Opportunities," *Technological Forecasting and Social Change* 11 (1977), 49–74 (see p. 54).

14. (New York, 1950), p. 205. A good deal of recent antiscientism is surveyed in Bernard Dixon, *What Is Science For?* (1973; rpt. Harmondsworth, 1976).

15. (New York, 1970).

16. J. D. N. Nabarro et al., "Selection of Patients for Haemodialysis," *British Medical Journal* 1 (March 11, 1967), 622–24 (see p. 623). Although several thousands of patients die in Britain each year from renal failure (there are about thirty new cases per million of population) only 10 percent of these can for the foreseeable future be accommodated with chronic hemodialysis. Kidney transplantation, itself a very tricky procedure, cannot make a more than minor contribution here. In the United States, about seven thousand patients with terminal uremia who could benefit from hemodialysis evolve yearly. As of mid-1968, some one thousand of these could be accommodated in existing hospital units and by June 1967, a worldwide total of some one hundred twenty patients were in treatment by home dialysis. Cf. R. A. Baillod et al., "Overnight Hemodialysis in the Home," *Proceedings of the European Dialysis and Transplant Association* 6 (1965), 99 ff.

17. For the Hippocratic oath see *Hippocrates: Works,* Loeb ed. (London, 1959), I, 298.

18. For various aspects of these issues see C. Doyle, "Spare-Part Heart Surgeons Worried by Their Success," *Observer* (London), May 12, 1968; J. Fletcher, *Morals and Medicine* (London, 1955); and G. E. W. Wolstenholme and M. O'Connor, eds., *Ethics in Medical Progress* (London, 1966).

19. Shana Alexander, "They Decide Who Lives, Who Dies," *Life* 53 (November 9, 1962), 102–25 (see p. 107). Cf. H. M. Schmeck, Jr., "Panel Holds Life-or-Death Vote in Allotting of Artificial Kidney," *New York Times,* May 6, 1962, pp. 1, 83.

20. A doctrinaire utilitarian would presumably be willing to withdraw a continuing mode of lifesaving therapy such as hemodialysis from a patient to make room for a more promising candidate who came to view at a later stage and who could not otherwise be accommodated. I should be unwilling to adopt this course, partly on grounds of utility (with a view to avoiding demoralizing insecurity), partly on the nonutilitarian ground that a moral commitment has been made and must be honored despite considerations of utility.

21. Lawrence Lader, "Who Has the Right To Live?" *Good Housekeeping,* January, 1968, pp. 85 and 144–50 (see p. 144).

22. In saying that past services should be counted "on grounds of equity" rather than "on grounds of utility," I take the view that even if this utilitarian defense could somehow be shown to be fallacious, I should still be prepared to maintain the propriety of taking services rendered into account. The position does not rest on a utilitarian basis and so would not collapse with the removal of such a basis.

23. In his contribution to Nabarro, "Selection of Patients," F. M. Parsons writes: "But other forms of selecting patients [distinct from first come, first served] are suspect in my view if they imply evaluation of man by man. What criteria could be used? Who could justify a claim that the life of a mayor would be more valuable than that of the humblest citizen of his borough? Whatever we may think as individuals none of us is indispensable." But having just set out this hard-line view he immediately backs away from it: "On the other hand, to assume that there was little to choose between Alexander Fleming and Adolf Hitler . . . would be nonsense, and we should be naive if we were to pretend that we could not be influenced by their achievements and characters if we had to choose between the two of them. Whether we like it or not we cannot escape the fact that this kind of selection for long-term hemodialysis will be required until very large sums of money become available for equipment and services [so that everyone who needs treatment can be accommodated]."

24. The relative fundamentality of these principles is, however, a substantially disputed issue.

25. This, of course, still leaves open the question of whether the point of view provides a valid basis of action: Why base one's actions upon moral principles? Or, to put it bluntly, Why be moral? The present paper is, however, hardly the place to grapple with so fundamental an issue, which has been canvassed in the literature of philosophical ethics since Plato.

26. Nabarro, "Selection of Patients," p. 622.

27. See Alexander, "They Decide Who Lives."

28. Nabarro, "Selection of Patients," p. 624. Another contributor writes in the same symposium, that "the selection of the few [to receive hemodialysis] is proving very difficult—a true 'Doctor's Dilemma'—for almost everybody would agree that this must be a medical decision, preferably reached by consultation among colleagues" (F. M. Parsons, p. 623).

29. Ibid., p. 623.

30. Dr. Wilson's article concludes with the perplexing suggestion— wildly beside the point given the structure of the situation at issue—that "the final decision will be made by the patient." But this contention is only marginally more ludicrous than Dr. Parson's contention that in selecting patients for hemodialysis "gainful employment in a well chosen occupation is necessary to achieve the best results" since "only the minority wish to live on charity" (ibid.).

31. To say this is, of course, not to deny that such questions of applied medical ethics will invariably involve a host of medical considerations. It is only to insist that extramedical considerations will also invariably be at issue.

32. M. A. Wilson, in Nabarro, "Selection of Patients," p. 624.

33. In the case of an ongoing treatment involving complex procedures, and dietary and other mode-of-life restrictions (and hemodialysis definitely falls into this category) the patient's psychological makeup, his willpower to "stick with it" in the face of substantial discouragements will obviously also be a substantial factor. The man who gives up, takes not his life alone, but (figuratively speaking) also that of the person he replaced in the treatment schedule.

34. To say that acceptable solutions can range over broad limits is not to say that there are no limits at all. It is an obviously intriguing and fundamental problem to raise the question of the factors that set these limits. This complex issue cannot be dealt with adequately here. Suffice it to say that considerations regarding precedent and people's expectations, factors of social utility, and matters of fairness and sense of justice all come into play.

35. Its historical credentials are impressive. Recourse to chance in this general connection runs deep in human thought. We speak of a person's

"lot" in life and find philosophers taking this idea seriously as early as the Myth of Er in Book X of Plato's *Republic*. Cf. also the classic treatise *Of the Nature and Use of Lots: A Treatise Historical and Theological* by Thomas Gataker (1574–1654).

36. One writer has mooted the suggestion that "perhaps the right thing to do, difficult as it may be to accept, is to select [for hemodialysis] from among the medically and psychologically qualified patients on a strictly random basis" (S. Gorovitz, "Ethics and the Allocation of Medical Resources," *Medical Research Engineering* 5 [1966], 5–7 [see p. 7]). To be sure, strictly random selection has the merit of affording a purely mechanical procedure that relieves its administrators of any difficult choices. But outright random selection would seem indefensible because of its refusal to give weight to considerations which, under the circumstances, deserve to be given weight (despite their departure from facile egalitarianism) on grounds not only of ethical considerations but of social advantage and the interests of the community at large. The proposed procedure of superimposing a certain degree of randomness upon the rational-choice criteria would seem to combine the advantages of the two without importing the worst defects of either.

37. Nabarro, "Selection of Patients," p. 623. The question of whether a patient for chronic treatment should ever be terminated from the program (say if he contracts cancer) poses a variety of difficult ethical problems with which we need not at present concern ourselves. But compare note 20 above.

38. *Statistical Abstract of the United States: 1974* (Washington, D.C., 1974), p. 62.

39. Data from B. C. Cooper, N. L. Worthington, and P. A. Piro, "National Health Expenditures: 1929–1973," *Social Security Bulletin,* February 1974, pp. 32–34.

40. Ibid.

41. Ibid.

42. John Pawles, "On the Limitations of Modern Medicine," *Science, Medicine and Man* 1 (1973), 1–30 (see p. 21).

43. *Genes, Dreams, and Realities* (New York, 1970), pp. 217–19.

44. Note that there are also cognitive ideals relating to the organization of knowledge (e.g., simplicity, consistency, completeness).

45. For some suggestions along these lines see my *Distributive Justice* (New York, 1965). It was the failure of the Pareto Principle in the context of the theory of justice offered in this book that motivated the deliberations of the present essay.

46. To say that the difficulties of an interpersonal comparison of utilities

need not faze the moral philosopher is not, of course, to indicate how he should deal with them. The writer's views on this theme are sketched in "A Philosopher Looks at Welfare Economics" in his *Essays in Philosophical Analysis* (Pittsburgh, 1969).

47. For a fuller development of the relevant themes see my *Unselfishness: The Role of the Vicarious Affects in Moral Philosophy and Social Theory* (Pittsburgh, 1975).

48. William Murdoch and Joseph Connell, "All About Ecology," *Center Magazine* 3 (Jan.–Feb. 1970), 56–63 (see p. 61).

49. Writing ca. 1700, the botanist John Ray observed that if the fossil evidence were taken at face value "it would follow that many species of shell-fish are lost out of the world," a conclusion which "philosophers hitherto have been unwilling to admit, esteeming the destruction of any one species as a dismembering of the Universe, and rendering it imperfect." (Quoted in John Passmore's excellent monograph, *Man's Responsibility for Nature* (London, 1977), p. 118.

50. In fact, Aquinas taught that Divine Providence is oriented with respect to lower creatures only to the species level; with intelligent creatures alone does God care for the welfare of individuals. *Summa contra gentiles,* 3.113.

51. The present focus on animal species is largely immaterial. Most of the present discussion applies mutatis mutandis to vegetable species and indeed to species of inanimate substance as well.

52. This point is elegantly formulated in Holmes Rolston III, "Is There an Ecological Ethic?" *Ethics* 85 (1974-75), 93–107: "For maximum noetic development, man requires environmental exuberance. . . . Remove eagles from the sky and we will suffer a spiritual loss. For every landscape, there is an inscape; mental and environmental horizons reciprocate" (p. 105).

53. On the hoary but nevertheless unfortunate tradition of ignoring — and even denying — the intrinsic value of nonhuman entities see James A. Keller, "Types of Matrices for Ecological Concern," *Zygon* 61 (1971), 197–209.

54. Some of us, to be sure, *are* so parochial. The utilitarians, for example, were generally committed to the view that "the creation of things, and the preservation of species, are to be aimed at and commended only in so far as human beings are, or will be emotionally and sentimentally [and intellectually and materially] interested in the things created and preserved" (Stuart Hampshire, *Morality and Pessimism* [Cambridge, 1972], pp. 3–4). And this is the mainstream position in Western philosophy, held alike by the Stoics in antiquity and the schoolmen of medieval Christianity: "Intellectual creatures are so governed by God as to be objects of care for their own sakes, but other creatures are, as it were, subordinated to the rational creatures" (Aquinas, *Summa contra gentiles* 3.112). Aquinas, however, is sufficiently astute to

add: "It is not contradictory for some creatures to exist for the sake of the intellectual ones *and also* for the sake of the perfection of the universe" (ibid.).

55. This discussion is somewhat incomplete. The argument it develops is directed against the "value-naturalism" of those who maintain that all value arises from interest. (See Ralph Barton Perry, *Realms of Value* [Cambridge, Mass., 1954.) Now the argument maintains the implausibility of this when interest is construed as specifically human interest. This leaves it open to a defender of the doctrine to insist that the scope of the interests at issue should be broadened to include that of certain other sentient beings, or perhaps even all of them, including the subhuman and conceivably the superhuman as well. At this stage, the counterargument would have to be broadened as well. But the reader can work the needed adaptations out alone.

56. See especially chap. 11 of Bk. 5 (Delta), and chap. 8 of Bk. 9 (Theta).

57. The atomists, who saw value as a specifically human creation, now come into their own and more.

58. Quoted in Arthur O. Lovejoy, *The Great Chain of Being* (Cambridge, Mass., 1936), p. 77.

59. For details, see the discussion in Nicholas Rescher, *Studies in Modality,* American Philosophical Quarterly Monograph no. 8 (Oxford, 1974), pp. 57–70. And cf. also the discussion of uniformity and diversity in Passmore, *Man's Responsibility for Nature,* pp. 119 ff. Aldo Leopold's seminal essay, "The Land Ethic" may count as an important venture in evaluative metaphysics. It is reprinted in *A Sand County Almanac* (New York: Oxford University Press, 1949), pp. 201–06. He maintains that: "A thing is right when it tends to preserve the integrity, stability, and beauty of the biotic community. It is wrong when it tends otherwise."

60. Historically, the traditional formulation of this line of thought proceeded in a theological idiom. Thus Aquinas says that all created things participate in the goodness of God, their creator (*Summa contra gentiles* 3. 9), so that everything "imitates the divine goodness, according to its own mode" (ibid. 3. 20), though, to be sure, in varying degree – "elements, for instance, [are of lesser goodness] than animals and man because they are unable to reach the perfection of knowledge and understanding, to which animals and men attain" (ibid.).

61. Passmore, *Man's Responsibility for Nature,* pp. 123–24.

62. *Morality and Pessimism* (Cambridge, 1972), pp. 36–37.

63. Of course, whenever such an "ought" of cosmic fitness is operative, there is a correlative duty to cultivate and promote its realization. But this represents quite a different issue. People ought to speak correctly or do their sums properly. But that does not make departures from correct speech or correct arithmetic *ethical* transgressions.

64. Here we revert to the issue (already mooted in note 55 above) of whether the individuals at issue are specifically persons or whether sentient beings in general are at issue. Moral philosophers have always been in discord in this question. Even within the utilitarian school Bentham (and later G. E. Moore) took the more liberal line while the Mills were more rigorously intellectualistic.

65. It should be obvious, but may nevertheless be worth stressing, that to say that saving species is not a moral duty is not to deny that various moral duties lie in this general area. It is clear that we have a moral duty to avoid the needless taking of animal life, the wanton destruction of animals or plants, the vandalization of our natural environment, and so on.

66. The issue is ably treated in Passmore, *Man's Responsibility for Nature*, chap. 5, pp. 100–26. And this discussion should perhaps be amplified by invoking the idea of a potential mutuality of recognition as "kindred spirits" that cannot readily be projected beyond the confines of the human community.

67. René Dubos has suggested extending the Decalogue with an eleventh commandment, "Thou shalt strive for environmental quality" (*A God Within* [New York, 1977], pp. 166–67). Two considerations arise. (1) Why should the limiting qualification "environmental" be present here at all? (2) In any event, the addition of this commandment transforms the Decalogue. It is now no longer a list of specifically moral injunctions but is expanded to reach into the wider ethical sphere, where we have left behind the parochial limitation to human interests as being paramount.

68. John Passmore holds much the same view in *Man's Responsibility for Nature:* "But that men have lost rights over them does nothing to convert animals into bearers of rights, any more than we give rights to a river by withdrawing somebody's right to pollute it" (p. 116).

69. *Sermons Preached on Various Occasions,* 2nd ed. (London, 1858), sermon 6, pp. 106–07. On the wider issues cf. Lynn White, Jr., "The Historical Roots of Our Ecological Crisis," *Science* 155 (1967), 1203–07.

70. One cannot treat a caterpillar unjustly. But to say this is neither to deny that one can abuse it (e.g., by utterly pointless destruction) nor even that one can be ungenerous toward it (e.g. by idly flicking it off a leaf it happens to be eating).

71. This also seems to be the position of Jacques Maritain. See his *Neuf leçons sur les notions premières de la philosophie morale* (Paris, 1951), as given in Charles Journet, *The Meaning of Evil,* trans. Michael Barry (London, 1963), pp. 138–39. For the denial of animal rights see also D. G. Ritchie, *Natural Rights* (London, 1894), p. 107; and John Passmore, "The Treatment of Animals," *Journal of the History of Ideas* 36 (1975), 195–218. For a

statement of the contrary position see Jay M. Lowry, "Natural Rights: Men and Animals," *The Southwestern Journal of Philosophy* 6 (1975), 109–22.

72. See the discussion of hedonic versus aristic values in the author's book *Welfare* (Pittsburgh, 1972), pp. 160–64.

73. The supportive argument at issue here is readily developed in deontic logic. Thus let:

$$\text{"X has a right to do A"} = \diamondsquare (\text{X do A})$$

We may then reason as follows:

"X does not have
 a right to do A" $= \sim \diamondsquare (\text{X do A})$
 $\leftrightarrow \sim \sim \boxed{\text{M}} \sim (\text{X do A})$ [Since: $\diamondsuit = \sim \square \sim$.]
 $\leftrightarrow \boxed{\text{M}} \sim (\text{X do A}) = \text{"X has an obligation not}$
 to do A"

The long and short of it is that one should not do that which one does not have a right to do.

74. This sentence is drafted with care not to read "the fact that we have no right to endanger another species entails that we have a duty to avoid threatening its existence." For this cognate, active rather than passively voiced version would not do the job which needs to be done at this stage of the argument.

75. *Man's Responsibility for Nature,* p. 101.

76. To be sure, various purely practical constraints are at work here. The attempt to destroy a dangerous species of bacteria or viruses might itself produce still more dangerous mutants.

77. It is moreover, at odds with preservationism itself. In endeavoring to save an endangered species we are, after all, interfering in the evolutionary course of events to produce in actual intervention a result that otherwise would not be.

78. Charles Hartshorne, "The Environmental Results of Technology," in *Philosophy and Environmental Crisis,* ed. W. T. Blackstone (Athens, Ga., 1974). pp. 69–78 (see p. 72).

79. "There is much to be said for the small group. It can work quite efficiently. Efficiency does not increase proportionally with numbers. A large group creates complicated administrative problems, and much effort is spent in organization." Laura Fermi, *Atoms in the Family: My Life with Enrico Fermi* (Chicago, 1954). p. 185.

80. Karl R. Popper in *Problems of Scientific Revolution,* ed. R. Harré (Oxford, 1975). p. 84.

81. Bernard Dixon, *What is Science For?* (1973; rpt. Harmondsworth, 1976), p. 143.

82. For a clear illustration of this phenomenon in chemistry, see F. H. Wertheimer et al., *Chemistry: Opportunities and Needs* (Washington, D.C., 1965), p. 9.

83. D. A. Bromley et al., *Physics in Perspective: Student Edition* (Washington, D.C., 1973), p. 16.

84. Use of the term *phenomena* and insistence upon their primacy in scientific inquiry goes back to the ancient idea of "saving the phenomena." See Pierre Duhem, *To Save the Phenomena,* trans. E. Doland and C. Maschler (Chicago, 1969); and Jürgen Mittelstrass, *Die Rettung der Phänomene* (Berlin, 1962). It finds its modern articulation in Newton's *Principia.*

85. See T. S. Kuhn, "Historical Structure of Scientific Discovery," *Science* 136 (1962), 760–64.

86. Generally, but not inevitably. For example, the main anomaly resolved by general relativity (the issue of the perihelion of Mercury) had been long familiar. Nevertheless a good deal of high-powered observational work using very sophisticated technology had to be done before general relativity became "established." The technological imperative for "new data" in natural science is usually operative on the side of the genesis of new theories, but even more emphatically so on the side of their validation.

87. See Bromley, *Physics in Perspective,* p. 331.

88. Thus in science, as in war, what matters is "gettin' thar fustest with the mostest." As his father wrote to the mathematician Johann Bolyai in 1823: "If you have really succeeded in the question, it is right that no time be lost in making it public. . . . every scientific struggle is just a serious war, in which I cannot say when peace will arrive. Thus we ought to conquer when we are able, since the advantage is always to the first comer." Quoted in Roberto Bonola, *Non-Euclidean Geometry,* trans. H. S. Carslaw, 2nd ed. (La Salle, 1938), p. 99.

89. Gerald Holton, "Models for Understanding the Growth and Excellence of Scientific Research," in *Excellence and Leadership in a Democracy,* ed. Stephen R. Graubard and Gerald Holton (New York, 1962), pp. 94–131 (see p. 115).

90. This dichotomization of problem areas finds its counterpart in the organization and management of scientific activities. The great national laboratories, institutes, and installations that typify big science reflect the concentration of effort needed to resolve power-intensive problems; the massive proliferation of other scientific activities among myriad smaller units (universities, industrial laboratories, etc.) reflects the decentralization of

effort to which the installment-plan aspect of complexity-intensive problems readily lends itself. For an interesting discussion of the relevant issues see Alvin R. Weinberg. "Institutions and Strategies in the Planning of Research," *Minerva* 12 (1974), 8–17.

91. For some interesting theoretical discussions of the concept of complexity and an extensive bibliography of the concept, see C. C. Whyle, A. G. Wilson, and D. Wilson, eds., *Hierarchical Structures* (New York, 1969).

92. Here a cousin to the myth of Sisyphus is applicable, that of the giant who rolls up the mountainside path a stone that becomes (nontrivially) heavier with each advancing step; be he ever so strong, even his most intense efforts will ultimately receive a check.

93. Reverting to the myth of the preceding footnote, observe that there need be no limit to how far the stone could be moved by the giant if it (and its pieces) could be split into halves at any stage.

94. The idea that "everything will be different" after the next breakthrough is tempting, but is clearly not a very scientific way of defending the interests of science. Scientific rationality demands adoption of a methodological uniformitarianism between the seen and the unseen. And we should surely use the same ground rules of explanation and method in discussions *about* science that apply in discussions *within* science.

95. On this line of thought compare the author's *Welfare: The Social Issues in Philosophical Perspective* (Pittsburgh, 1972), pp. 155–66.

96. The moral aspect aside, it is probably not even a matter of prudentially enlightened self-interest. Cf. the cogent discussion in Stephen Toulmin, "Is There a Limit to Scientific Growth?" *Scientific Journal* 2 (1966), 80–85.

97. See Keith Norris and John Vaisey, *The Economics of Research and Technology* (London, 1973), p. 56.

98. Some of the issues raised here are dealt with more fully in the author's *Scientific Progress* (Pittsburgh, 1978).

Name Index

119

Subject Index

121